Mari Ruti and Climate Change

In this illuminating book, Clint Burnham invites the reader to consider humanity's relationship with the world around us, using a unique blend of Lacanian psychoanalysis, literary criticism, and visual art to hold a mirror up to our own implications in the mounting climate crisis.

Drawing upon the pioneering work of philosopher Mari Ruti, Burnham deftly interweaves examples from climate fiction – including works from Richard Power, Eleanor Catton, and Jenni Fagan – 'trash art', and classic films from Alfred Hitchcock to help the reader explore the idea of and better understand what is now called climate grief. Focusing on sublimation and creativity, Burnham weighs up perspectives on both climate activism and climate denialism and uses these ideas to offer a form of respite from trauma or grief surrounding the climate crisis, providing both comfort and a bracing call to action.

Mari Ruti and Climate Change offers a novel and approachable perspective to both students and scholars interested in psychology, environmental studies, psychoanalysis, and climate politics, as well as practitioners of the psychological and therapeutic professions who are encountering patients experiencing climate anxiety or other affects in their practice.

Clint Burnham is a professor of English at Simon Fraser University, Canada. His research spans psychoanalysis, film studies, cultural studies, and literary theory.

Essays in Psychological Humanities
Series Editor: Matthew Clemente and David Goodman

The *Essays in Psychological Humanities* Book Series is home to short, timely titles authored by today's leading thinkers on the most pressing and perennial problems we face. Written for academics, practitioners, and educated readers alike, books in this series attempt to address the fundamental human questions of suffering, meaning, mortality, love, and potential – issues of life and death that concern us at the core. Literary in style, philosophical in nature, offering the depth, wisdom, and psychological insight that every inquiry into the human condition aims to uncover, these works are essential reading for thinking persons today and promise to provoke thought and enliven conversation for years to come.

Mari Ruti and Climate Change: From Grief to Creativity
Clint Burnham

For more information about this series, please visit: https://www. routledge.com/Essays in Psychological Humanities/book-series/EPH

"In this haunting and personal study of the politics of climate grief, Clint Burnham draws on the work of the late Mari Ruti to create a unique cartography of contemporary creativity, desire, and loss. At a moment of climate catastrophe, Burnham's brilliant, genre-defying book provides us with the intellectual resources we need to live with grief without surrendering to despair."

Imre Szeman, *Professor of Human Geography and Director of the Institute for Environment, Conservation and Sustainability at University of Toronto-Scarborough College, author of* Solarities: Seeking Climate Justice *and* On Petrocultures: Globalization Culture and Energy

"Burnham deftly weaves biography (his own and that of Mari Ruti), climate fiction, environmental world events, and Ruti's psychoanalytic theory to reveal what is so pressingly needed today: an emancipatory theory of (climate) grief, in which the personal and the eco-political are dialecticized and creatively sublimated. A tour de force."

Ilan Kapoor, *Professor, Faculty of Environmental and Urban Change, York University, Toronto, co-author of* Global Libidinal Economy *(2023)*

"Love as critique, fidelity as betrayal. In this unabashed meditation, or essaying of/for the other, Clint Burnham bears witness to the force and provocation of Mari Ruti, to her immeasurable impact on his thought and life. We are treated to an irresistibly unruly work. Meanings of grief and trauma multiply; the personal chases and exceeds the universal. If 'clusterfucks' are the new normal, Burnham gifts us a work for the moment."

Zahi Zalloua, *Cushing Eels Professor of Philosophy and Literature, Professor of Indigeneity, Race, and Ethnicity Studies, Whitman College, USA, author of* Being Posthuman: Ontologies of the Future *and* Žižek on Race: Toward an Anti-Racist Future

"In *Mari Ruti and Climate Change*, Burnham brings together climate change, auto-theory and Lacanian ethics in a dialectical clusterfuck of grief which considers how we deal with the polycrises, both contingent and universal, of the present. Drawing on Mari Ruti's philosophy and, in particular, her late work on sublimation, Burnham contends that grief and enjoyment (jouissance) might intermingle in dangerous ways. Cli-fi, Hitchcock films, forest fires and freedom convoys all operate as sites of stackable, fungible grieving which bundle together with personal grief – of lost fathers, cancerous bodies and relationship breakdowns. At stake in *Mari Ruti and Climate Change* is a question of ethical subjectivity and how sublimation might well be a risk worth taking, and even enjoying, towards emancipation."

Rosemary Overell, *Senior Lecturer in Media, Film &*
Communication at the University of Otago, New Zealand,
editor of Post-truth and the Mediation of Reality:
New Conjunctures *and author of* Fisting the Dead

"Our ecological crisis demands new thinking and Clint Burnham's *Mari Ruti and Climate Change* offers us the necessary tools. By drawing on Mari Ruti's original Lacanian development with a perspective vital precisely where it diverges from Slavoj Žižek and the Ljubljana school, Burnham demonstrates the transformative potential of sublimation and creativity over the destructiveness of the drives and the Real."

Michael Gray, *Host of the Žižek & So On podcast*

"This book is a beautiful exposition of grief through a Lacanian lens with the memory of the late philosopher Mari Ruti (1964–2023) in mind. It is an intimate questioning of her writings on the theory of sublimation, progressive social theory, the ethical subject and, ultimately, the question of being when there is no cure. Burnham is also interested in the climate crises and the fungibility of grief, whether it stacks, and what art and fiction might offer to our understanding of climate denialism. The question of temporality, desire, and what it means to know something about one's pending death accompany the reader through Burnham's masterpiece of a book."

Sheila L. Cavanagh, *Professor of Sociology, York*
University, author of Queering Bathrooms:
Gender, Sexuality, and the Hygienic Imagination *and*
Sexing the Teacher: School Sex Scandals and Queer Pedagogies

"Reader, can't you see that we're burning? This is the question that scorches us throughout this book. The nauseating, tragic odour of burning trees/forest fires infuses the text, we think of ashes (to ashes), the smell of smoke (while we are sleeping), the burning fossils (in Powers' *Overstory*); in short Burnham doesn't take the heat off from the get-go. (The Lacanian in me cannot resist hearing in "Clint Burnham" the imperative "Clint, burn 'em!" henceforth.) The burning (of the planet) is contemplated through readings of culture, climate change theories and activism, and questioned through Lacan's notion of an ethics of desire in the painstaking, stone-turning reading of (what Burnham calls) Mari Ruti's 'politics of grief'. Burnham's friendship with Mari Ruti, and his respect for and engagement with her work is no hindrance here to his passion and skill in bringing out the paradoxes and contradictions between her early and later writings on lack and sublimation to lead us to a brilliantly articulated and innovative contemplation of climate grief."

Dr Carol Owens, *Psychoanalyst, Dublin, and Editor at PCSReview (Psychoanalysis, Culture and Society)*

"By way of his creative engagement with the work of the late Mari Ruti, Clint Burnham expertly turns on its head what we thought we knew not only about climate grief but also about loss itself. In a hauntingly beautiful style, what emerges is a new dialectical politics of grief based on a grief as divided and contradictory as we who grapple with it. Burnham asks, what does the fact of our ontological lack mean for the possibilities of grieving, and of whether creativity helps us to sublimate as a response to ecological grief and trauma? Through Ruti, Lacan, climate fiction, art, and the importance of style, methodology, and critique, this book is not only a fitting homage to Ruti but also one in which Burnham's voice undeniably intervenes with something new."

Stephanie Swales, *Assistant Professor of Psychotherapy, Dublin City University, author of* Perversion: A Lacanian Psychoanalytic Approach to the Subject

"Clint Burnham's *Mari Ruti and Climate Change* is an absolute must for anyone who wants to understand Mari Ruti's thought. Burnham's book makes clear the urgency of Ruti's theory of sublimation as a way of dealing with trauma through his investigation into climate grief. He bravely tackles the nature of grief in the face of multiple traumas. Through Ruti's work, as well as Lacan and Hegel, the book proposes a new dialectics of grief, one which does not seek easy solutions but rather engages the relationship between contingent lacks and our constitutive lack. Through fascinating readings of psychoanalysis, the fires in Canada, and Hitchcock's *Vertigo* (just to name a few), Burnham not only honors the singularity of Ruti but demonstrates how to respond meaningfully to our current moment."

Hilary Neroni, *Associate Professor of Film and Television Studies, University of Vermont, author of* Feminist Film Theory and "Cléo from 5 to 7" *and* The Subject of Torture: Psychoanalysis and Biopolitics in Television and Film

Mari Ruti and Climate Change

From Grief to Creativity

Clint Burnham

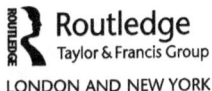

Routledge
Taylor & Francis Group

LONDON AND NEW YORK

Designed cover image: Mike Surber. The painting within the image is
Escaping Criticism by Pere Borrell del Caso.

First published 2026
by Routledge
4 Park Square, Milton Park, Abingdon, Oxon OX14 4RN

and by Routledge
605 Third Avenue, New York, NY 10158

*Routledge is an imprint of the Taylor & Francis Group, an
informa business*

British Library Cataloguing-in-Publication Data
A catalogue record for this book is available from the British
Library

ISBN: 978-1-032-85594-3 (hbk)
ISBN: 978-1-032-80703-4 (pbk)
ISBN: 978-1-003-51891-4 (ebk)

DOI: 10.4324/9781003518914

Typeset in Times New Roman
by Apex CoVantage, LLC

This book is in memory of Mari Ruti, of my father, Lee Burnham, and for those suffering from wildfires, through death, dispossession, and displacement.

Contents

Acknowledgments

Mari Ruti and Climate Change began as a series of talks in the fall and winter of 2023–2024, at Otago University in New Zealand (online), at the Mahindra Seminar at Harvard, for Lacan in Scotland (online), and at the Theory Institute of the Society for Critical Exchange at Wakulla Springs, Florida. Thank you first to Rosemary Overell, Frances Restuccia, Calum Neill, and Jeffrey R. Di Leo for their generous hospitality. Thank you as well to the various interlocutors who listened to and commented on these early versions, including Sheldon George, Matthew Flisfeder, Kelly Gray, Matthew Mersky, Paul Kingsbury, Scott Krzych, Kirk Boyle, Jake McDonald, Lucas Pohl, Russell Sbriglia, Cindy Zeiher, Zahi Zalloua, Cary Wolfe, Derek Woods, Aaron Jaffe, Nicole Simek, and Jane Gallop. Early ideas for the book were germinated at the Facing Ecological Grief workshop at Simon Fraser University's Faculty for the Environment in spring 2023: thank you to Naomi Krogman, Dean, and Paul Kingsbury for the kind invitation. Versions of Chapters 3 and 4 were also presented at the MLA in New Orleans, Louisiana, in 2025 and at the LACK conference in Columbus, Ohio, that same year. Thank you as well to Eva Graham for inviting me into their class to talk about *The Sunlight Pilgrims* and to Matt Hern and Am Johal for visiting my classroom to talk about their book *Global Warming and the Sweetness of Life: A Tar Sands Tale*. Working with first Zoe Meyer and then Deepika Batra and Ayushi Awasthi at Routledge has been a pleasure: thank you to Zoe for the commission at the Psychology and the Other conference in Boston and to Deepika for the deadline. Gratitude for the shared memories of Mari from Alice Kaplan, Jill Gentile, Hilary Neroni, Heather Jessup, Judy Hamilton, Todd McGowan, Ryan Engley, Gail Newman, and Kathryn Kuitenbrouwer. For all that Lacanians cling to the Real and the death drive, they're actually quite nice people, and I want to mention especially how important two collectives have been for me: the Lacan Salon (2007–2023: Paul Kingsbury, Hilda Fernandez, Alois Sieben, Alma Krillic, Ted Byrne, Alessandra Capperdoni), and the dysUnited group (2023–2025: Risa Mandell, Rana Sioufi, Joseph Scalia, Todd Dean). I also want to thank my students (Eva Graham, Ziwei Yan, Jake McDonald, Rawia Inaim, Alois Sieben, Deanna Fong, Miyona Katayama, Nicholas Collard, Bobby Malone, Sofia Shah, Lakeisha Barrington, Kanksha Chawla, and Odessa Twibill in particular) and colleagues (Cornel Bogle, Joanne Leow, Jeff Derksen, Steve Collis, Deanna Reder, Sophie McCall, David Chariandy, Margaret Linley, Carolyn Lesjak, Mike Everton, Troy Sebastian) at Simon Fraser University for their insight and collegiality. Conversations with Jamie Hilder, Alberto Toscano, Larissa Lai, Scott Innis, Michael Turner, Tim Lee, Chris Brayshaw, Cecily Nicholson, Mercedes Eng, Tania Willard, Nate Crompton, Emily Fedoruk, Danielle LaFrance, Rolf Mauer, Hamish Ballantyne, Tara Bigdeli, Rita Wong, and Kim Trainor were indispensable. The index was prepared by Jake McDonald (thanks again) and funded by Simon Fraser University's University Publication Fund, for which grateful acknowledgment and thank you are expressed, especially to Steeve Mongrain, who helped me fill out a particularly gnarly online signature form – the sinthome indeed. One of Mari's tricks was knowing when and how to retreat to get the work done, and this book took many stints at the Pender Island cabin of Norah and Ken Sawatsky (for which many thanks) including a digital detox – another Ruti hack. And thank you and much love to my partner Julie Sawatsky, and our offspring Devon, for their support, conversations during hikes on Vancouver's North Shore mountains (on the unceded traditional territories of the səlilwətaɬ – the Tsleil-Waututh Nation, Sḵwx̱wú7mesh Úxwumixw – the Squamish Nation, and xʷməθkʷəy̓əm – the Musqueam Nation), and for keeping me in stitches. *Mari Ruti and Climate Change* was written during the genocide in Gaza: to misquote a particularly Lacanian passage in Adorno's *Aesthetic Theory*, this book is defined in relation to what it does not contain.

Abbreviations of Mari Ruti's works referred to in the text

"BB": "The Brokenness of Being: Lacanian Theory and Benchmark Traumas." *Angelaki* 28.6 (Nov. 2023): 123–70.

D: *Distillations: Theory, Ethics, Affect*. New York: Bloomsbury, 2018.

EOO: *The Ethics of Opting Out: Queer Theory's Defiant Subjects*. New York: Columbia UP, 2017.

SB: *The Singularity of Being: Lacan and the Mortal Within*. New York: Fordham UP, 2012.

"WCNC": "When the Cure Is that There Is No Cure: Melancholia, Mourning, Creativity." *Meaningless Suffering: Traumatic Marginalisation and Ethical Responsibility*. Eds. David Goodman and Mookie Manalili. New York: Routledge, 2024: 4–28.

Introduction

A bunch of us met to have dinner that night, but I left and walked off by myself, bought the silver ring, a bag of chips, then sat in the main square and bummed a cigarette off an old French man, then continued to sit there for many hours until the man with the bulgy eyes came to sit next to me and flirt. . . . Perhaps it's true what they say about the planet heating up – I mean, it's quite obviously true . . . Perhaps this is the true reason I do not finish books. Perhaps this is what has been blocking me.

– Sheila Heti, *Alphabetical Diaries* 3, 148, 149

There's nothing rational about grief – maybe you're crying for yourself.
– Marianne Jean-Baptiste in Mike Leigh's
Secrets and Lies (1996)

Autobiographical

This book is the culmination of my thinking in a psychoanalytic way about climate that goes back, at least in my work, to an intervention at the Petrocultures conference in Edmonton, Canada, in 2012, when I was talking about Edward Burtynsky's photographs of tailing ponds and pipelines and the tar sands projects in Fort McMurray. There I criticized the conception of the arboreal forest where the tar sands are located as some pure or unsullied nature, the lack in the big Other that we mistake for a lost object. I think I was interested in Burtynsky's photographs not only because of scale (a small tailings pond, because presented in a large photograph, looks more impressive/depressing) or their artworld connotations (an aerial photograph of rectangular tar sands ponds resembles a Mark Rothko or Barnett Newman abstraction) but also because, in the case of a photograph of pipelines near Cold

DOI: 10.4324/9781003518914-1

Lake Alberta, I had lived there (at the Canadian Air Force base in Cold Lake) in the 1970s when my father was an air traffic controller. And so I knew, or felt, that since the area was the largest air weapons testing range in North America, that this wasn't some pristine Nature with a capital N. This was a militarized, colonized space. More recently, the Vancouver Lacan Salon organized a conference on Lacan and the environment in 2018, which was followed by a collection of essays of the same title edited by Paul Kingsbury and me and published in 2021 by Palgrave – the conference, and the book, collected some important work by Cindy Zeiher, Todd McGowan, Nathan Gorelick, Hilda Fernandez, among others. In our introduction, Kingsbury and I discussed the famous Lacan anecdote of the sardine can floating in the ocean as a figure of nature, but also pollution, as the lacking Other. We were also interested in the figure of the climate denialist, in part because of how Kingsbury has worked on proximate subjects like UFO enthusiasts and Bigfoot hunters, those who similarly mistrust science but also seem to want to be scientists (or, as another similar subject, the anti-vaxxer, will say: do your own research!).

Then, in the spring and summer of 2023, a number of events impinged upon my work in a more immediate, not to say contingent, fashion. First, shortly after she spoke at the LACK conference in Burlington, Vermont, the psychoanalytic theorist Mari Ruti died of breast cancer. This was a blow for many of us, and already my chronology is mixed up, because between seeing Mari at LACK in mid-April and hearing of her death two months later, I spoke at a conference in Vancouver called "Facing Ecological Grief." Little did I know that summer 2023 would turn out to be the worst summer in history, here in Canada, for forest fires: more hectares burned by many times over (already in May), smoke enveloping the east coast of the United States and floating over to Europe, towns evacuated, firefighters recruited from Australia and South Africa and France, firefighters dying. Where I live, in the city of Vancouver on Canada's west coast (on, as we say, the unceded traditional territories of the Coast Salish peoples, including those of the Squamish [Sḵwx̱wú7mesh Úxwumixw], Tsleil-Waututh [səlilw̓ətaʔɬ], Musqueam [xʷməθkʷəy̓əm], and Kwikwetlem [kʷikʷəƛ̓əm] Nations), the smoke was sometimes present, indeed was so when I began writing what became this book. There is something about waking up in the middle of the night and smelling smoke, breathing it in, to bring on all kinds of affects. Finally, in August 2023, my own father died, which while it happened suddenly – he had dementia for the previous two years – in some ways he was already not there, between two deaths, as we say in the jargon. I had "pre-grieved" as a character says on the TV show *Succession*. The climate crisis, the deaths of two people close to me: to which grief should I attend? Did I have a choice?

But you want more here. I had known Mari for almost a decade, from when we were keynote speakers at a psychoanalytic conference in Toronto in the fall of 2013. I taught her book *The Singularity of Being* in a graduate course in the spring of 2016: here, I thought, was a way of thinking about Lacan and "the Real" that was beholding neither to humanist ideas of the subject nor to the newer "death drive" oriented readings coming out of Žižek, Todd McGowan, and Lee Edelman. I taught all of this work in the context of trans theory and poetics (the work of Trish Salah) and Lacan's seminar on the *sinthome* (*Seminar XXIII*) to a very small cohort (I believe my graduate chair had to convince our dean to allow a course with fewer than five students to run: three of the students, Deanna Fong, Alois Sieben, and Nick Tan, went on to write brilliant PhD dissertations). Mari and I met again at a Lacanian conference (the first in the "LACK" series) in Colorado Springs in the spring of 2016, where Mari, again, was a keynote speaker. I was not, but I did pitch a project I was then working on for a book series she edited at Bloomsbury. (Introducing Mari for her keynote, Todd McGowan gave a fulsome speech that credited Mari's book *The Summons of Love* with helping him woo his wife, Hilary Neroni. Mari promptly demolished that conceit, pointing out that her book was published in 2011, when Todd and Hilary had already been married for more than a decade.) The following summer, she was a featured speaker at an "Affiliated Psychoanalytic Workgroups" conference "On Love" in Vancouver. At this point, I had already put her name down for my promotion to full professor at my university – the process itself is so arcane and bureaucratic that while I suspect she was a referee, I could never be certain. We were close enough colleagues that, during her summer visit to Vancouver, we hiked the "Grouse Grind," a steep mountain trail much frequented by outdoor enthusiasts, but also not close enough so that when she invited me back to her apartment for a drink afterward, I knew that this was merely a polite gesture, one that I could decline. Our relationship continued in this fashion, meeting at conferences every year or two, until, in the fall of 2021, we were both attending an online conference "in" Boston, and she gave a talk, "When the cure is that there is no cure: mourning, melancholia, and creativity," during which she spoke in a matter-of-fact way about her terminal cancer diagnosis, and how the theories of Lacan, Freud, and Kristeva helped her cope. This was a great shock to many of us who, while we knew Mari and her work, were not in her inner circle – and the distance, social and otherwise, brought about by the pandemic had not helped anything. I emailed Mari a couple of times. Word spread that a colleague was fundraising for medical expenses – even the much-vaunted Canadian public health-care system had, it seemed, failed Mari. I contributed. When I saw her again in person in spring 2023, we were sitting at the same cafeteria table in a New England university town, and I did not recognize her.

Forest fires that year were the worst ever:

Canada's 2023 wildfire season was the most destructive ever recorded. By the end of the year, more than 6,000 fires had torched a staggering 15 million hectares of land. To put that in perspective, that's an area larger than England and more than double the 1989 record. Normally, an average of 2.5 million hectares of land are consumed in Canada every year. And unlike previous years, the fires this year were widespread, from the West Coast to the Atlantic provinces, and the North. By mid-July, there were 29 mega-fires, each exceeding 100,000 hectares.

("Canada's record-breaking wildfires")

The fire season started early in May and June of that year (although so-called zombie fires will last for years, dormant and underground [Shingler]), and smoke from the fires spread across North America (Broadway shows were canceled) and to Europe. My BC Wildfire app tracks fires across British Columbia, Canada's most western province, and where I live, in the southwestern coastal city of Vancouver: when I consulted it while on a camping trip in August 2023, the red dots for out-of-control fires covered the entire map on my phone; even when I pinched and grabbed the screen to enlarge the image, there seemed to never arrive a space without some fire's icon – yellow for being held, green for under control, or the sigil of flames in a red circle, for a "Wildfire of note." Those were the ones you had to worry about – more so even than fires that were out of control, fires that, in spite of dozens or hundreds of men and women on the ground, helicopters or water bombers in the air, millions of dollars being spent ("The annual national cost of wildland fire protection exceeded $1 billion per year for the past decade" ["Cost of Wildland Fire"]). The magnitude of

Canadian fires shown by more of such statistics. Supernatural sunsets due to the particulates in atmosphere are not simply from woodfire, as the plastics and other synthetic materials go up in smoke. Breathing difficulties and advisories for the young, old, or respiratory-afflicted to avoid the outdoors reversed the recent protocols we had learned during the pandemic. And there is something about smelling smoke when you are sleeping to bring on all kinds of affects.

That August of 2023 I was camping with my family in the Slocan Valley in the eastern portion of British Columbia. Called the "Kootenays," the region is the territory of the Ktunaxa First Nation. An Indigenous writer of the Ktunaxa, nupqu ʔak·ɬam Troy Sebastian, also wrote about those fires, offering a global perspective in his essay "From the ashes of my father's house: a memoir from COP28." nupqu ʔak·ɬam is not interested in clichés of melancholy and loss but instead the details of a local fire, the "St. Mary's fire," from that summer:

> Since the fire the power was cut off, refrigerators and freezers are now biohazards. They sit outside of each surviving home awaiting removal – resembling children dropped off at the wrong bus stop. The community's power lines have been removed to ensure no further sparks create havoc and devastation. Some lines burned, others stayed in place, like empty sentinels upon a vista of black charred ground. Toothpick thin pine and poplar are scorched to death, yet still standing.
>
> (Sebastian)

In comparison, my own misadventures when camping are negligible: was I ever in real danger? And it wasn't like I was in the backcountry with only a rucksack and water I filtered out of a local stream. Truth be told, we were "glamping," staying in a small peak-roofed cabin that offered snug beds for three. But, of course, such abodes are hardly protection. That summer I was reading John Vaillant's *Fire Weather*, an exhaustively researched account of the 2016 fires in Fort McMurray in Northern Alberta. Among the surreal details – propane bottles popping as the fire destroyed hundreds of homes in the epicenter of the Canadian oil industry, footage from nanny cams in houses that had been abandoned to the fires – Vaillant contrasted two historical laboratories for determining how fires can destroy human habitation. "We don't have a forest fire problem, we have a home ignition problem," he quotes Roy Rasker, "cofounder of Community Planning Assistance for Wildfire"

(Vaillant 144), before going on to describe a 2005 experiment by Underwriters Laboratory that showed how midcentury furniture burned in comparison to "modern equivalents." The contemporary living room went up in three minutes, in what is called a "flashover" (145) due to the large amount of "petroleum products and related chemicals" used in today's buildings and furniture. A similar experiment, with much more lethal intent, was conducted by the Allied forces during World War II, when British and American scientists built models of German houses. This was to determine the two-step process of high explosives, which would first demolish or partially demolish buildings, knocking out windows and doorways. Then incendiary bombs, "containing phosporous and thermite" (Vaillant 206) would land "in beds or on the floor beside wardrobes, chest of drawers, behind bedframes" as one survivor recalled (Miller 126, qtd. Vaillant 206).

For all of this – the doomer statistics, the on-point analysis – there is a certain amount of enjoyment to the forest fires, especially for those on the environmental left. Now we are vindicated (but we cannot be too smug: a small-town official in Alberta who in 2016 tweeted "Karmic #climatechange fire burns CDN oilsands city" was suspended from his position [Maclean]) and so our grief is mixed in with *jouissance* in a different way than the grief of mourning the death of a colleague or friend. But I also found these emotions conflicting with each other: I had not done grieving for Mari as the forest fires escalated, but as I already noted, my chronology was misplaced, grief resists such neat ordering, for while I had been thinking about ecological or climate grief specifically after seeing Mari in April but before her death in June, this question of how to sort these different traumas had preoccupied my mind at least for a couple of years, since giving a talk on the "clusterfuck" of 2020 in an online conference in the fall of that year. There I referred to the pandemic, anti-Black state violence, and the climate disaster as different claimants on our attention (Burnham, "*Nil Actum Credens*"). And it's no good consulting the old Scots proverb cited by Trollope in *Barchester Towers*, "It's gude to be off with the auld luve/ Before ye be on wi' the new" (Trollope 241, 448). The character this is addressed to, Obadiah Slope, tries to woo two women at once:

> Of the wisdom of this maxim Mr. Slope was ignorant, and accordingly, having written his letter to Mrs. Bold, he proceeded to call upon the Signora Neroni. Indeed, it was hard to say which was the old love and which the new, Mr. Slope having been smitten with

both so nearly at the same time. Perhaps he thought it not amiss to have two strings to his bow. But two strings to Cupid's bow are always dangerous to him on whose behalf they are to be used. A man should remember that between two stools he may fall to the ground.

(Trollope 241)

These maxims – concerning the old and new love, or the two stools – suggest that our rhetorical understanding of emotions or the subjects that occasion them have a certain fungibility to them, and this was further torqued in my case with my father's death in August 2023. I was already finding the emotional interplay of climate and mourning to be difficult. But my father's death, an overdetermined Freudianism notwithstanding, was all the more hazardous. Without subjecting the reader to too much detail of years of arguments, disappointments, friction, and drama,[1] I will just focus on the scene where I learned he had died.

My family (there remain four of us: two sisters and two brothers; our mother died in 1992) knew this was coming, and my father had been in a long-term care facility for just over a year, after a rapid decline into dementia brought on, perhaps, or at least accelerated, by the anomie of the pandemic lockdown. I last saw him alive in May 2023, and at least the Alzheimer's had tempered his propensity for verbal nastiness. He was at last a nice guy. The second week of August, my sister Charlene, who was in close contact with the nursing home, said he had suffered a rapid decline and was now confined to bed, not expected to rise again. He had a "Do Not Resuscitate" – a DNR. When, some seven or eight years earlier, he appointed Charlene and me as his coexecutors with power of attorney, and he told us, at his lawyer's office, he wanted a DNR, I joked that if he stubbed his toe, that was it. But this time it was for real. He lived some four hours away from my brother Greg and me, and so we made plans the following Saturday (this was a Tuesday or Wednesday) to drive up, a trip that also involved a two-hour ferry ride. We had just arrived on the ferry on a Saturday night and walked up from the car deck to sit in the passenger lounge when my phone rang. Charlene was telling me our father was dead. I had a crappy connection, a loudspeaker was making safety announcements, and for some reason, as I walked through the lounge to try to find a quiet space, I was surrounded by raucous fans of the BC Lions football team, in their orange jerseys, the color rhyming in a weird way with the sunset to the west. We were late, my brother and I, and all my emotions erupted in anger at stupid football fans, idiotic safety protocols, and moronic cellphone providers and their inadequate service. I tried to tell myself, this was probably later, undoubtedly later, that, well, my dad was a football fan (I've never cared for sports, one of many differences) and maybe he would have liked the synchronicity.

Outline of argument

And so here is the argument I am making *in nuce*: the question of the fungibility of grief, of whether grief stacks, or we deal with one then another, or they make each other worse, is the most important question in the contemporary politics of grief. Furthermore, Mari Ruti's work on grief and sublimation – her essays "The Brokenness of Being" (hereafter "BB") and "When the Cure Is that There Is No Cure" (hereafter "WCNC") – is crucial to this understanding. We begin with the clusterfuck of the current decade's eco-political landscape: climate, anti-Black violence, COVID-19, the rise of populism and late fascism, intractable wars in the Sudan, Gaza, and Ukraine. Then, in 2023, the terms of my own origin story for writing this book entail theorizing the fungibility or prioritizing how to grieve Mari's death, the summer of forest fires, and my father's death, providing an orthogonal dimension. In both instances – the clusterfuck and the singular – the question is how these griefs or lacks or traumas relate to one another. Does one replace the other, can we only encounter one at a time, or do they make each other worse, or does one lack perhaps help us deal with another? Ruti's late theory argues for creative sublimation as a response to/articulation of contingent or constitutive or benchmark trauma versus ontological lack on the one hand and socioeconomic on the other. That is, just as the historical present – the clusterfuck – is met by the subject in terms of the fungibility of grief, Ruti's positing of creative sublimation is also, or already, bifurcated in terms of how it can alleviate, in differing forms of efficacy, ontological or contingent trauma as opposed to the socioeconomic. Reading these essays as creative responses (and thus capable of being interpreted in a figural fashion) but also in the context of Ruti's earlier work and their origins in psychoanalysis and philosophy tells us that this question of creativity and sublimation (what Lacan memorably describes as "raising an object to the dignity of a Thing") finds idiosyncratic expression in Ruti: jewelry, Cézanne, matchboxes, Lacan's *das Ding*. Like Sheila Heti in the epigraph, she buys a silver ring, and it somehow consoles her. In each case, the example or object is met by its opposite: is buying jewelry a way of staving off consumerism or caving into it? Can we compare one of the great modernist artists to someone placing matchboxes on their mantlepiece? Those readings in Ruti and Lacan are then extended or challenged via other cultural objects, especially climate fiction, where creativity as a response to ecological grief is a figure for art itself, and garbage art, where the artwork made of garbage stands in for an impossible synthesis. Those readings in turn contribute to a robust psychoanalytic theory by confronting key moments in the Freudo-Lacanian tradition. *Antigone* and the dialectic of desire teaches us, in Lacan's reading, that "not giving ground relative to one's desire" is easier said than done when our "desire is the desire of the Other." Freud's kettle logic and four "*Ver-*"s, and Kübler-Ross's stages of grieving, all allow ways of taxonomizing our contradictory responses to climate change. Melville's Bartleby and Hegel's Beautiful Soul, again, offer the fantasy of preferring not to engage and enjoying one's virtue signaling. Circling back to foundational or ontological lack, and its relation to loss (and the loss of loss) and to the lack in the big Other, means that since there is no big Other, nature is not lost but, like climate change, itself lacks, you can't make up with your own lack, "and no sacrifice can compensate for this lack of the [big] Other" (Žižek, *Enjoy Your Symptom* 55).

To work out these questions of creativity and sublimation with respect to climate grief, I now turn to two cli-fi novels, *The Sunlight Pilgrims* and *The Box*, and a sculpture, *Saturación 00*. Jenni Fagan's *The Sunlight Pilgrims* immediately poses a classificatory dilemma if we read it as a creative response to the trauma of climate change, however, for two reasons: first, set in the winter of 2020 (although published in 2016), it raises the false impression that it will deal with the COVID-19 pandemic; second, it concerns not global warming but a sudden deep freeze: in the penultimate chapter, we are told the temperature is −56°C:

> The landscape is brilliantly lit, flawless – the mountains look like somebody has cut them out of the sky. The skies are clear and blue, but the wind still bites and nips at any exposed inch of skin. Each of them wears snow goggles so their eyelashes don't get frosty, and balaclavas pulled up right up over their face and nose. The cold is clangorous. It vibrates. Shrill and deadly.
>
> (Fagan, chapter 38)

Mandy-Suzanne Wong's *The Box* (2023) also poses problems for the cli-fi category, not only in the same substantive way as *The Sunlight Pilgrims* (it concerns a brutal and unending winter laying waste to a city) but also for how, again in both novels, we have creativity as a response to grief by way of refusing, or foreclosing, the very object of that grief. Here is what I mean: in *The Sunlight Pilgrims*, the very question of what I earlier called the "clusterfuck" or the polycrisis (climate change, anti-Black racism, the pandemic, the rise of populism, wars in Ukraine and Palestine) is operationalized when a character is repeatedly queried as to whether they are more worried about the brutal winter or transphobia; in *The Box*, climate denialism appears as intergenerational violence – that is, it is displaced onto one's elders as a way to disavow one's own agency. These creative responses then are matched by sublimation, Lacan's elevating an object to the dignity of the Thing, in Adán Vallecillo's *Saturación 00* (2017), which not only renders oil filters into works of art but in so doing transforms the traces of pollution into an aesthetic.

In Jenni Fagan's *The Sunlight Pilgrims*, we are in a town in Scotland during an unprecedented cold winter: two of the central characters are Constance, who is hastily sketched as akin to a disaster "prepper," and her child, Stella, who is transitioning. These two characterological types, or actants, show us how in the novel there is a dual or struggle for primacy between climate change and trans issues.

First, in chapter 19, we see Stella's account of dealing with change room issues (not having access to a gender-free space, and the transphobic classmates):

> Stella can't even explain how much she has dreaded gym class after that. It's worth going through an Ice Age just to not have to do that again. She could lie down in the snow like an angel and wait for winter to take her home.
>
> (Fagan, chapter 19)

Then, in chapter 23, in an email chat with Vito, an Italian trans friend, he asks:

> *What's the weather like there? Are you scared it's an Ice Age?* Stella finishes combing her hair while she thinks about it. She types a reply. *More scared about how to go through transition, don't know how to do it. I don't want any operations either, not even when I'm older.*

For Stella, issues of transphobia and how to transition – access to puberty blockers, and surgery – are more urgent (less scary, worth going through) than climate change. In the novel, the extra-cold winter stands as a dialectical inversion or negation of the "actually existing" conditions of global warming – the old name for climate change.

The intergenerational conflict (here a matter of my interpretation, but in the novel writ large, that between the survivalist mother and trans child) is also operative in *The Box*, when we learn that acts of violence in an unnamed city's unending winter have been carried out by a group of students in retribution to (my) generation's (why not, I'll take the blame) malfeasance with regard to climate change:

> "*They* ravished the Earth so it's gone cuckoo and forgotten how to administer climates and meteorological routines; consequently there's so much snow we can't ride our electric scooters and shan't ever have cars; they've . . . disconnected the electricity, so we can't plug in our phones and stream pornography and purchase useless paraphernalia over the internet; and now they're threatening electricity rationing because of the climate emergency . . ." The old woman heard them opine that everyone over thirty should be put out of the way unless they're providing for children.
>
> (Wong 179)

We have, in the two novels' "creative" responses to climate grief, to the climate crisis, then, both the question of how to prioritize or swap out one crisis or grief for another (transphobia versus climate change) and then the disavowal of one's responsibility for anthropogenic climate change by way of intergenerational conflict. Vallecillo's *Saturación 00* is a 15- or 20-foot-high U-shaped sculpture made of recycled car filter paper. I saw the work at a survey exhibition at the ICA Boston in the fall of 2023: *Forecast Form: Art in the Caribbean Diaspora 1990s-Today* was curated by Carla Acevedo Yates and originated at the MCA in Chicago. The bottom of the sculpture rested on the floor, a gesture that pushed back against a simplistic reading of the work. That is, initially, we see the object as a form, one in dialogue with the minimalism that characterized late modernist art in the 1960s and 1970s. But then, upon perusing the didactic panel, we learn that, indeed, the work is made from recycled car filters: now we understand that the work has in effect been made by a combination of the impurities that are filtered out from a car or truck's engine oil, and the structure of the filter paper itself, folded or pleated so as to form a tube in the filter. That is, there is now an antagonism between the formal qualities of the work and the poetics of its materials, with the latter bequeathing an environmental sensibility that also problematizes simply glorifying the artist as maker. But the sculpture's resting of its bottom "U" on the gallery floor then works against both understandings. Now the viewer must stand back from the work and cannot lean in and examine it closely (or only can from the side, awkwardly). The gallerygoer is reminded of their body and that their body and the sculpture itself have a relationship with each other.

But insofar as dirt and pollution are elevated here into an artwork, *Saturación 00* enters into conversation with *The Sunlight Pilgrims* and *The Box*. The novels and the sculpture help us understand the question I attack again and again in this book: the question of whether creativity help us to sublimate as a response to grief, to trauma, a raising of the object to the dignity of the Thing and then the problem of whether privileging trans issues over climate, shifting blame for anthropogenic climate change, or negating the reader's or viewer's relation with the artwork constitute forms of climate denialism *which then turn out to be a fidelity to one's idiosyncratic desire as rooted in one's ontological lack.*

Chapter breakdown

This book derives its argument from two conditions: first, Mari Ruti's two late-in-life essays "The Brokenness of Being" and "When the Cure Is that There Is No Cure" constitute a manifesto for the role of sublimation as a way of dealing with trauma, which logic I then extend to climate grief itself; second, the situation of Ruti's own death as a way of considering the fungibility of grief or how we prioritize (or even can prioritize) our lacks or trauma. In the first chapter I examine the essays themselves closely, paying attention to their textuality and rhetorical figures, endeavoring to read them as closely as Ruti reads Lacan or Kristeva. This occasions thinking about Ruti's turn to auto-theory but also the signal cultural example of sublimation she draws from Lacan: the still-life paintings of Cézanne. I extend that reading via a consideration of Hitchcock's film *Vertigo*, another *nature morte*, perhaps (the French term for the still-life genre) in its presentation of California's redwood forest. The chapter culminates with the first of a few semiotic rectangles, graphs that, like Lacan's various algorithms, seek to offer ways to understand the conceptual.

But I also want to discuss Ruti's work more broadly, both as a way to situate her final essays in that grander sweep and also to push at some of her theoretical arguments more rigorously. In the second chapter I take on three texts, primarily *The Ethics of Opting Out* (hereafter *EOO*), *Distillations* (*D*), and *The Singularity of Being* (*SB*). These all date from the 2010s and abjure any autotheoretical ramblings, instead making strong arguments with respect to a *rapprochement* between affect theory and psychoanalysis, and, especially, for the importance of what we can call "progressive critical theory," a reading I supplement with a consideration of three novels from the same time period and also so concerned – Jeffrey Eugenides's *The Marriage Plot*, Dionne Brand's *Theory*, and Michelle de Kretser's *Theory and Practice*. I also delve more deeply into a key component of Lacan's reading of the *Antigone*, which Ruti has argued troubles, or perhaps dialecticizes, the question of desire. If one must not cede one's desire, as Lacan tells us, how does that jibe with the precept that desire is the desire of the other? What, then, is the ethical subject? The chapter concludes with Ruti's reading of Lacan's *sinthome*.

In the third chapter, I engage more thoroughly with, on the one hand, so-called cli-fi or climate fiction – considered both in terms of Amitav Ghosh's influential pamphlet *The Great Derangement* (2017) and Richard Powers's Pulitzer Prize–winning novel *The Overstory*. Plot spoiler: I don't actually like *The Overstory* but rather draw on one specific moment in it (an environmentalist character's life story is stolen by a government agent, with disastrous outcomes) to argue against Ruti's sublimation theory. But I also posit against Powers's depiction of climate activism an "actually existing" example from the "war in the woods" on Canada's west coast. Those activists are in turn juxtaposed with the portraits of oil workers in Matt Hern and Am Johal's *Global Warming and the Sweetness of Life: A Tar Sands Tale* as a way of testing out the question of whether Freud's four forms of negation – denial, repression, foreclosure, and disavowal – can help us understand a psychoanalysis of climate denialism. The chapter concludes with a discussion of Tita Salina's *1001st Island*, an artwork about climate disasterism that is interpreted with the aid of a semiotic rectangle. (Freud's "kettle logic" will make an appearance in the following chapter.)

I begin the fourth chapter continuing a discussion of that semiotic rectangle and continuing to explore the psychoanalysis of climate change via Žižek's adaptation of Elisabeth Kübler-Ross's five stages of grieving, repurposed for climate grief. This account of how we think about, or do not think about, or suffer from, climate change is then contrasted with the more scientific analysis exemplified in such visualizations as the graphs of the Great Acceleration, so-called dick pics of climate scientists. But the method followed throughout this book – dialectics that bring the different together – also cannot neglect questioning its own methodological tools. So, I take on a verity of Lacanian theory, the disavowal of Hegel's "Beautiful Soul," who is seen as too hesitant to dirty their hands in politics or activism, and the valorization of Melville's Bartleby, who "would prefer not to." In effect, I seek to bring the first up a bit and bring the second down a bit, in terms of thinking about how we, all of us (here the universal returns), are implicated in the climate crisis. But this autocritique cannot neglect Ruti's work itself – nothing is more dreary than the critic too much in love with their topic. (I may have had a crush on Ruti as a person but can think of no more loving gesture than critique.) I conclude the chapter by thinking about Ruti's work in terms of Lacanian centrism and centrist Lacanianism (which are not the same thing).

In the book's coda, I draw on those notions of Lacanian centrism and centrist Lacanianism to understand the politics of the ethical subject (that is to say, is the climate denialist not refusing to cede ground to their desire?) and the fungibility of grief. I then finish with the question of Ruti's texts as a late style and argue that the idea of a writer's late style takes on a different relation to temporality when she has foreknowledge of her own death. In that regard, Mari may have known more than we can, although perhaps, by now, we can foresee a planetary diagnosis that we should face as unflinchingly as she did hers.

Note

1 An obituary I wrote can be found here: https://www.comoxvalleyrecord.com/obituaries/morley-bruce-lee-burnham-3104311.

References

Brand, Dionne. *Theory*. Toronto: Penguin Random House, 2018.

Burnham, Clint. "*Nil actum credens, si quid superesset agendum*: Or, Slavoj, Can't You See I'm Burning? Žižek *avec* the Clusterfuck of 2020." *Understanding Žižek, Understanding Modernism*. Eds. Jeffrey Di Leo and Zahi Zalloua. London: Bloomsbury, 2022. 77–89.

Burnham, Clint. "Photography from Benjamin to Žižek, via the Petro-chemical Sublime of Edward Burtynsky." *Proceedings of the Petrocultures Conference (See Conference Papers, Below)*. McGill-Queen's UP.

Burnham, Clint and Paul Kingsbury, Eds. *Lacan and the Environment*. London: Palgrave, 2021.

"Canada's Record-Breaking Wildfires in 2023: A Fiery Wake-up Call." *Natural Resources Canada*, 27 December 2024. https://natural-resources.canada.ca/stories/simply-science/canada-s-record-breaking-wildfires-2023-fiery-wake-call. Accessed 11 May 2025.

"Cost of Wildland Fire Protection." *Natural Resources Canada*, 15 January 2025. https://natural-resources.canada.ca/climate-change/climate-change-impacts-forests/cost-fire-protection. Accessed 12 May 2025.

de Kretser, Michelle. *Theory and Practice*. New York: Catapult, 2024.

Eugenides, Jeffrey. *The Marriage Plot*. New York: Fourth Estate, 2011.

Fagan, Jenni. *The Sunlight Pilgrims*. Hogarth, 2016, ebook.

Ghosh, Amitav. *The Great Derangement: Climate Change and the Unthinkable*. Chicago: U of Chicago P, 2017.

Hern, Matt and Am Johal. *Global Warming and the Sweetness of Life: A Tar Sands Tale*. Cambridge: MIT P, 2018.

Heti, Sheila. *Alphabetical Diaries*. Toronto: Penguin Random House, 2024.

Maclean, Rachel. "Tom Moffatt Feels the Heat Over 'Karmic' Fort McMurray Fire Tweet." *CBC News*, 5 May 2016. https://www.cbc.ca/news/canada/calgary/tom-moffatt-karmic-tweet-fort-mcmurray-fire-1.3568341. Accessed 12 May 2025.

Miller, Carl F., Ed. *Fire Fighting Operations in Hamburg, Germany During World Warr II*. Washington: Civil Defence Preparedness Agency, 1971.

Ruti, Mari. "The Brokenness of Being: Lacanian Theory and Benchmark Traumas." *Angelaki* 28.6 (November 2023): 123–170.

Ruti, Mari. *The Case for Falling in Love: Why We Can't Master the Madness of Love – and Why That's the Best Part*. Naperville, IL: Sourcebooks, 2011.

Ruti, Mari. *Distillations: Theory, Ethics, Affect*. New York: Bloomsbury, 2018.

Ruti, Mari. *The Ethics of Opting Out: Queer Theory's Defiant Subjects*. New York: Columbia UP, 2017.

Ruti, Mari. *The Singularity of Being: Lacan and the Mortal Within*. New York: Fordham UP, 2012.

Ruti, Mari. *The Summons of Love*. New York: Columbia UP, 2011.

Ruti, Mari. "When the Cure Is that There Is No Cure: Melancholia, Mourning, Creativity." *Meaningless Suffering: Traumatic Marginalisation and Ethical Responsibility*. Eds. David Goodman and Mookie Manalili. New York: Routledge, 2024. 4–28.

Sebastian, nupqu ʔakᐧɬaṁ Troy. "From the Ashes of My Father's House: A Memoir from COP28." *IndigiNews*, 12 December 2023. https://indiginews.com/first-person/from-the-ashes-of-my-fathers-house-a-memoir-from-cop28.

Shingler, Benjamin. "It's the Middle of Winter, and More than 100 Wildfires Are Still Smouldering." *CBC News*, 21 February 2024. Accessed 12 May 2025.

Trollope, Anthony. *Barchester Towers*. Harmondsworth: Penguin, 1985.

Vaillant, John. *Fire Weather: The Making of a Beast*. Toronto: Knopf, 2023.

Wong, Mandy-Suzanne. *The Box*. Toronto: Anansi, 2023.

Žižek, Slavoj. *Enjoy Your Symptom! Jacques Lacan in Hollywood and Out*. New York: Routledge, 1992.

Chapter 1

Mari Ruti's work

Overview

This chapter thinks with Mari Ruti's late works "When the Cure Is that There is No Cure" and "The Brokenness of Being" (the first appears in Routledge's *Meaningless Suffering: Traumatic Marginalisation and Ethical Responsibility*; the second was published in *Angelaki*, with a commentary by Hilary Neroni), reading them to better understand what is now called climate grief. Ruti drew from these two texts in talks she gave at the *Psychology and the Other* conference in Boston in fall 2021 and at the LACK conference in April 2023, less than two months before her death from cancer on June 8 of that year, talks that are the culmination of her remarkable body of work over the past two decades but also mark a turn to "autotheory" as she grapples with her own diagnosis.

So let us begin with an overview of Ruti's oeuvre, followed by a discussion of the vexed problem of autotheory, particularly as it is circumscribed for us in Anna Kornbluh's new book *Immediacy*. Mari Ruti's work spans two decades of books, talks, essays, and interventions, and includes from my point of view three remarkable texts: *Distillations: Theory, Ethics, Affect*; *The Ethics of Opting Out: Queer Theory's Defiant Subjects*; and *The Singularity of Being: Lacan and the Mortal Within*. I know I am leaving out some important texts, including *A World of Fragile Things*, *The Case for Falling in Love*, *The Call of Character*, *Penis Envy and Other Bad Feelings*, the biography of Julia Kristeva written by Alice Jardine and edited by Ruti, and her book on Lacan and Melanie Klein, written with Amy Allen. But it was in *Distillations*, *The Singularity of Being*, and *The Ethics of Opting Out* that Ruti staked out a reading of

DOI: 10.4324/9781003518914-2

Lacan that is not only amenable to so-called affect theory and queer theory but, more importantly in my view, one that diverges from a certain tendency in post-Lacanian theory (the Slovenian school and its American followers), which focuses on the Real as some horrifying and traumatic encounter, "to privilege the destructiveness of the drive over the lures of desire" ("WCNC" 10). In *Distillations*, for example, Ruti argues that the new universalism (found in works by not only Žižek and Badiou but also such North American academics as Todd McGowan, Paul Eisenstein, Zahi Zalloua, and Ilan Kapoor) "does not adequately distinguish between various identitarian movements" precisely in the sense that they "take it for granted that every singularity can claim an immediate membership in the universal" (*D* 27). Nonetheless, Ruti is also able to make the implausible connection between the Lacanian universalists and affect theorists, as when she argues that Sara Ahmed's "feminist killjoy" evinces the kind of rupture Eisenstein and McGowan theorize, or Badiou's "possibility of the impossible" (*D* 49). I should stress that, for me, I am sympathetic to Ruti's critique as well as the universalist negativity of her opponents. My conjoining of these signifiers in an adjectival subordination of universalism to the negative should not signal a hierarchy, for they are both surely a legacy of the appropriation of Hegel first in French philosophy in the 1930s and 1940s and then its continuation via the Slovenians and their North American epigones – on the one hand, Lacan is the vanishing mediator (Butler 186–204), and on the other hand, it has been suggested that Badiou's role with respect to the universal has to do more with skirmishes in French politics in the 1990s, when he published his Saint Paul book.[1]

Ruti's laying out of the argument regarding the universal and the particular in *Distillations* has two implications for my larger argument with respect to a psychoanalytic approach to climate grief. First is whether climate change itself can be thought of in terms of the universal. There is a certain commonsense truth at work here, to the degree that the overheating of the planet, rising sea levels – that is to say, the very overabundance of climate disasterism, not to say the turn to the Anthropocene as a human-centered concept (I discuss this latter via its scientific visualization in Chapter 4) – but this is to neglect the problem of trauma, isolated, individual(ism) and Ruti's later arguments with theory. Here I should admit that I will swap out, fairly willy-nilly, trauma and grief in particular. There's no excuse for it: I know grief is our response to trauma, but as George Orwell observed decades ago with respect to the causality of suffering and alcohol, the one leads to another with little respect for niceties.

The theoretical stakes are perhaps evident in Lee Edelman's rejoinder to Ruti's critique, in his *Bad Education: Why Queer Theory Teaches Us Nothing*. Mari Ruti, he tells us, "repurposes liberal humanism, vaunting 'social survival, justice, and responsibility' without recognizing that the split in the subject, the fracture at the core of posthumanist thought, makes those three things incompatible" (Edelman 213). I may be unfair in characterizing Ruti and her opponents in this way; it's worth noting that she had exceedingly cordial relations with many of the figures I just mentioned; too, Anna Kornbluh goes out of her way to cite Ruti in her recent book, which cannot honestly be said to be describing Ruti's project in its characterization of autotheory, even if, generically, that is the self-described trajectory of Ruti's late work. I'll come back to Kornbluh's critique later and return to *Distillations* and *Opting Out* in the following chapter.

Ruti's reading of Lacan is well known and also perhaps most controversial, for its tendency to be misread as overly optimistic: thus, in "Zizekians/Lacanians: What do you think about Mari Ruti's The Singularity of Being?", a social media post (on the r/zizek subreddit), "jebemkodyodrana" argues that Ruti "tries to paint the Real as a positive force which helps the human escape her enclosed being in [the] symbolic order. [at] odds with the Real developed by Zizek and Zupančič, as something that is out of human control and doesn't have this characteristic."[2] In a slightly more rigorous register, Tom Ratekin draws on Ruti to argue in favor of identification with one's *sinthome*: "The most significant part of one's being is not nationality, race or even a quality that one has chosen but is instead the way one inexplicably experiences jouissance" (Ratekin 159).

Two essays and their tropes (semiotic rectangle 1)

To sit with "When the Cure Is that There is No Cure," and "The Brokenness of Being," what is important about the two texts is how they treat what Ruti refers to as the "idiosyncratic" or "contingent" lack that is a bout with cancer, or other forms of "benchmark trauma" that appear as if out of nowhere, in comparison with the "ontological" or "constitutive" lack that is the psychoanalytic incurability of human existence, as well as with the "systemic" or "structural" lack that is due to sociopolitical conditions – the injuries of racism, classism, cisheteropatriarchy, and/or colonialism. That is, she is dealing here with what we might see as the "clusterfuck" I have proposed as a way of thinking about climate change alongside anti-Black violence and COVID-19, or an overlapping "over-lack" of lacks, and how they stack up against one another. (Lee Edelman proposed a resistance to assigning questions of "urgency," to hierarchizing priority in dealing with the same issues, arguing that they are rather "ongoing aspects of human political experience, inevitable outcomes of the cleavage in the subject's relation to jouissance," although he does seem to make an exception for climate change [Edelman et al. 2018].) But this question of an overlap of grief or trauma, of its fungibility, is also one that turns up in novelistic (or "cli-fi") treatments (see Chapter 3), for the contemporary social novel today rarely treats the climate crisis separable from other forms of trauma or grief, even as, argued by Amitav Ghosh, it is both incapable of responding to said trauma and yet implicated in causing it (Ghosh 10, 23). Ruti's observation is that, in part, for her, when she received her cancer diagnosis in 2018 (and was told she had 12 to 18 months' life expectancy, so she beat those odds), she was better able to receive that bad news because she already had a Lacanian diagnosis baked into her psyche. So, she is positing not an ontologization of grief (as is frequent in the normative accounts found in Renée Lertzman [2015], Sally Weintrobe [2021], or Donna Orange [2017] – the best-known psychological authorities on climate grief[3]) but instead a dialectics thereof. Thus, whenever in this book I talk about the fungibility of grief that Ruti's work helps us discern, I am not so much arguing for grief as a *thing*, or ontology, but as a divided substance, one characterized by paradox and dialectics, by contradiction.

Ruti analyzes "benchmark traumas" via psychoanalytic theory for two reasons: first, following Lacan's argument that "human subjectivity *as such* [is] centred around the kind of ontological lack, wound, or injury for which there is absolutely no cure . . . the lack of a remedy that characterizes benchmark trauma is, for Lacan, built into the very constitution of subjectivity" ("BB" 132). Ruti's writing here is important, I believe, for its contradictions. One of these, for instance, lies at the heart of her "When the Cure Is that There is No Cure" thesis, and in the second reason she argues that psychoanalytic theory can be helpful:

Let me restate the argument as follows: if we were not fundamentally lacking . . . we would not have the creative, innovative capacity to endure the kinds of more contingent lacks (crisis situations, everyday disasters) that most of us, from time to time, are destined to experience. If we had no familiarity – however unconscious – with what it feels like to be lacking, we truly would not know how to react when something went terribly wrong in our lives; we would not know how to even begin to formulate a response to the calamity at hand. I have chosen Lacan, Klein, and Kristeva as my main interlocutors because all three grasp this reality: they share an understanding of the very relationship between constitutive lack, creativity, and our (always fragile) ability to handle contingent traumas that I wish to explore.

("WCNC" 5)

She phrases the question somewhat differently, but with the same effect, in "The Brokenness of Being":

My sense is that those who have gathered an awareness of the incurable nature of their ontological lack have often also developed an array of psychic and affective strategies for coping with benchmark traumas – strategies that may not be available to those who have avoided their lack or sought to fix it with easy solutions, such as rampant consumerism.

("BB" 166)

I don't know if I agree. It seems to me that contingent lacks, or "benchmark traumas," also have more force, indeed only have any force because of constitutive lack.[4] To reverse the Ruti thesis here is also to posit the coincidence of opposites. Perhaps it is the very coinciding of ontological lack and that which happens "once in a lifetime" or in an "unprecedented" way – the benchmark traumas of cancer, or anthropogenic climate change—that makes the latter so brutal. Or, rather, the reason we think of such lacks, correctly labeled by Ruti as contingent or idiosyncratic, is that their psychic effects can be, well, contingent. Some of us laugh it off, some of us descend into a trough of depression and melancholia. Some of us dwell on it for years, and some of us veer back and forth between denial and embrace. The two words "contingent" (used three times in the "WCNC") and "idiosyncratic" (used more often, almost two dozen times) are themselves telling.[5] Contingent is rarer an occurrence in Ruti's text, I think, because it is closer, connotatively speaking, to the socioeconomic – to precarious employment. We speak of contingent faculty in the academy – the professors' version of the precariat, what Raymond Williams called the intellectual proletariat. As we shall see, Anna Kornbluh suggests that this economic factor plays a role in the development of autotheory. And yet "contingent" or contingency also suggests a randomness, an aleatory quality: and here we can think of the clinical nostrum that an analysand's presenting symptom is rarely that which marks their lack – if a patient comes in and says that they are troubled by forest fires or rising sea levels, that this gives them anxiety, then the job of the clinic is to move past those utterances – the *sujet d'énoncé* or enunciating subject – and follow the network of signifiers to the fundamental fantasy, the cleavage in the jouissance – the *sujet d'énunciation* or the subject of the statement, the unconscious. Indeed, when the subject declares that their mother died of cancer two years ago and now a logged forest's devastation "brings it all back," or if I find myself unable to grieve my father's recent death because my "grief stack" is already filled up with worrying about the summer of 2023's forest fires, about breathing that smoke, the fungibility of this trauma, or grief, its contingency, to return to Ruti's keyword, should also give us pause. Perhaps the stackable isn't scalable to account for the idiosyncratic.[6] Then, "idiosyncratic" seems more individual, personal – the singularity that is a hallmark of Ruti's thinking. Etymologically, the word "idiosyncratic" carries the suggestion of blaming the victim ("attributable to individual disposition, susceptibility, or character" – *Oxford English Dictionary*) but also owing, as with "idiom," or "idiot," to the ancient Greek root ἴδιος, meaning a "private," or "lay," but also "uneducated," person. I confess that I do like the idiotic connotations of idiosyncratic: surely I was an idiot to go camping in the wilderness during Canada's worst forest fire season in history.

I want to look at this contradiction in more depth, and also other contradictions that I think are operative in Ruti's essays, as a memorial "working through" of my grief, which is to say my grief with Ruti's death that spring, which seems to then have accumulated or weaponized on the one hand, climate grief when faced with this summer of forest fires, and also a contingent or idiosyncratic grief of encountering my father's death at the end of August. It goes without saying – which is why I am saying it – that I am talking about the contradictions not as a way of pointing out inadequacies in the texts (which would be a real dick move, since they are by necessity posthumous works), but because the most important texts have contradictions at their heart.

Mari Ruti essays' full titles are as follows: "When the Cure Is that There Is No Cure: Melancholia, Mourning, Creativity" and "The

Brokenness of Being: Lacanian Theory and Benchmark Traumas."
So, with the first we have a direct Freudian reference – his "Mourning and Melancholia" essay – and a signal to what will be the most
important avenue when there is not a cure: creativity. Creativity is
what we turn to as a cure for there being no cure, as a way to ac-
cept. Ruti draws on Jacques Lacan, Julia Kristeva, and Melanie Klein,
as she tells us to see that "creativity, potentially at least, offers an
antidote to melancholia as an ontological (existential) predicament"
("WCNC" 1). Moreover, "creativity can serve as a productive – albeit
always incomplete and imperfect – response to both ontological lack
and benchmark trauma" ("BB" 133); and mourning (for Kristeva and
Freud) and melancholia (for Lacan and Klein) are themselves step-
ping stones to that creativity.

Such creativity, Ruti asserts in "Brokenness," does not or should
not be offered as a cure for systemic trauma – since such inequities
can presumably "be fixed by more just and egalitarian social arrangements." (She returns to the question of systemic or structural trauma
late in "WCNC.") Ruti hedges her bet. She recognizes that "collective
remedies to structural injustices are unlikely to materialize anytime
soon and that many dispossessed individuals are therefore left to their
own devices in trying to find ways to cope with their predicament"
("WCNC" 2). Later, she expands on the role of creativity in such conditions by way of the negative:

> Creativity should **not** be seen as an alternative to political and prac-
> tical efforts but merely as a supplement to such efforts – as a tem-
> porary placeholder for the sturdier remedies and solutions we hope
> to achieve. . . . [D]enying the potential of creative activities, which
> could be as simple as painting a rundown apartment or setting up
> a community garden in a poor neighborhood, to provide solace on
> grounds of ideological purity would be a mistake for the simple rea-
> son that whatever can be done to diminish people's suffering in the
> here and now is valuable and does not in principle need to interfere
> with our pursuit of long-term solutions.
>
> ("WCNC" 29)[7]

Such material-aesthetic actions marry the creative and the ameliorative and bear some family resemblance to the Black Panthers' "reform
without reformism" or "survival pending the revolution," as a history
of the party is titled (Alkebulan 2012). Or, to be more precise, their

supplementary "placeholder function" suggests a secondary role. But here I should also address what may be bothering some readers, or my placing of anthropogenic climate change in the category of the "benchmark trauma" or grief to which creativity can be a useful response (and not merely the ameliorative role suggested for structural inequities) but for which there is no cure. Is writing about or making art about (or making theory about) climate change what we do precisely because we cannot stop it? Or does it fall into the category of climate mitigation, which can include anodyne behaviors like swapping out air conditioners for heat pumps or the specter of decarbonization, and so on, for which it would be obscene to slap on the label of Black Panthers' slogans. This category question is not clear in Ruti's work, for she does note that "it can at times be difficult to differentiate between [benchmark] traumas and systemic, structural traumatization that results from unequal sociopolitical and economic arrangements" ("BB" 131), saying so immediately after making a catalog of what forms benchmark trauma can take: "[d]eath, or the anticipation of death" (her cancer diagnosis) and also debilitating accidents or illness, suicidal depression or anxiety, police violence, and joblessness. But I think this problem we have reminds us of a more existential, if I can put it that way, or perhaps ontological, which is to say political question of whether climate change is a trauma for which there is no cure (like Mari's cancer diagnosis) or that which can be "fixed by more just and egalitarian social arrangements" (as Ruti argues is the case for "structural traumas such as poverty, racism, sexism, or other social inequalities" ["WCNC" 1]).

I think Ruti's concept of creativity is more than simply a sublimation of the proper political action or agency into the aesthetic – such would be a model of sublimation more Freudian than Lacanian, after all, since for Lacan sublimation is not so much a diversion of the libidinal into the aesthetic (as per Freud – and of course Freud saw political action itself as a form of sublimation, indeed one necessary for human civilization); Lacanian sublimation is not that diversion, but rather, as he famously put it in *Seminar VII*, the seminar on the ethics of psychoanalysis, what he called the raising of an object to the dignity of the Thing (*Seminar VII* 112). In "Brokenness," Ruti develops this idea in more depth, arguing that the Thing is what is most terrifying and most desirable, both sublime object that destroys us and a "fantasy-infused placeholder for everything that we imagine having lost" ("BB" 133).

Here is Scott Krzych on how Ruti conceptualizes sublimation (pay attention to a certain Ruti keyword):

Ruti proposes, after Lacan, that only certain unpredictable objects will provide for a subject the kind of auratic experience essential to the Thing. As [Ruti] sees it, sublimation enacts a commitment to one's own idiosyncratic, but also singularly consistent, attitude to desire. Sublimation, then, is "a matter of struggling to find ways to incorporate the echo of the Thing (as the cause of our idiosyncratic desire) into the rhythm of our social lives even when doing so proves demanding."

(Krzych 131)

I should like to remark on Ruti's consistency – she remains faithful to this notion of idiosyncratic desire developed in *A World of Fragile Things* in "Brokenness of Being":

The sublimatory gesture of raising an ordinary object into an extraordinary status allows us to draw closer to the truth of our desire – the kind of desire that is governed by our jouissance rather than by the desire of the big Other, the dominant sociocultural and linguistic establishment that Lacan calls the *symbolic order* and that strives to dictate the contents of our desire.

("BB" 133)

Idiosyncratic desire, idiosyncratic lack – we might discuss Ruti's "incurable contradictions" in four sites or nodes of sublimation: one conceptual, that of creativity, and three having to do with objects that also, in her two texts, function as *topoi* or figures: the beach and ocean, the ring, and Cézanne's (but also iPhone's) apple.[8] I begin with one of these figures: the ring. Ruti tells us that "objects that contain a strong aura of the Thing can connect us to our jouissance," that "some objects have the power to lessen the sting of our ontological lack and of the melancholia that haunts so many of us due to this lack," even as she is "not looking to heal [her] wounds or to obtain ontological wholeness" ("WCNC" 11–12).

During a recent trip to a small town in the South of France, I bought two silver rings, meant to be worn together, one with a little white pearl and the second with seven narrow bands bunched together to create a wider band. I do not usually wear rings, and rarely buy items during trips, but these rings somehow leapt at me from the display case. The psychoanalyst Marion Milner notes how finding an elegant dress once made her feel recentered and re-empowered after a vexing social interaction that had made her feel personally erased. I would not go that far. But I cherished the rings, and they quickly became associated with the magical month that I spent in a town that I loved and that gave me a break from the morose realities of being ill. The rings even became associated with the fact that after my trip, I received the first encouraging CT and MRI scans in three and a half years: my cancer had stabilized.

("WCNC" 12)

Commenting on an episode where she loses and then finds the rings, Ruti says "It is clear that, for me, these rings carry a poignant aura of the Thing. And the fact is that I do not care that I know full well that their appeal is entirely fantasmatic" ("WCNC" 12). So we have, first, the question of how Lacan's sublimation (raising the object to the dignity of the Thing) is brought into service for coping with grief (in Ruti's case, the contingent grief of cancer, which we can extend, perhaps, to climate grief) but also inquire just where that Thing is, if it is in fact external to the subject, or perhaps internal. The rings as a signifier is, like the apple, an overdetermined trope. Also, Ruti's utterance that she "know[s] full well" the "appeal [of the rings] is entirely fantasmic" carries a suggestion of Mannoni's *je sais bien . . . mais quand même*, or the formula of fetishist disavowal that is the hallmark of the pervert's *Verleugnung*.

But Ruti, as I said, discusses the ring as *das Ding* in both texts: she gives a different origin story in "Brokenness," telling us that

In the context of spending six weeks in Vienna, I found three relatively cheap silver rings in a store that sold trinkets. They immediately caught my eye. I felt summoned by them. I bought two of them right away and came back for the third a few days later because I could not get it out of my mind. The point of the story is that I know, on a deep level, that these are the only rings I will ever want for the rest of my life. I may wear others that I already own, and I am not going to reject a ring that someone might give me as a present, but I know with absolute certainly that I myself will not buy any new rings because these three fully satisfy my desire.

("BB" 163)

Two origin stories both have to do not only with rings as sublimated objects but also a kind of knowledge, an "absolute certainty." And Ruti goes on to declare, in "Brokenness," that the rings' status as objects of her desire, as Things that allow her to articulate her idiosyncratic desire, allowed her to escape the clutches of consumerism and capitalism. This is another role of sublimation, alongside that of coping with grief – be it benchmark or ontological. And it also helps us to distinguish – and this is really the "inside baseball" moment of this chapter, the Lacanian rabbit hole – the Thing from the *objet petit a*. The *objet petit a* was developed by Lacan around the same time as the nonce-concept of *das Ding* and refers not so much to a thing that we have sublimated but instead to the very succession of things from which Ruti wished her rings protected her.

Calum Neill's recent handbook *Jacques Lacan: The Basics* offers a refresher on this distinction. Desire, Neill points out, is rarely for one object or achievement: "Very few people who write a book or a song, or paint, only write one book or one song, or paint one picture. Fewer people only have sex once" (55). (These examples work well because they also have to do with creativity.) The *objet petit a*, Neill adds, then is not so much a placeholder for that succession of objects or achievements, but rather the empirical objects are a substitute for *objet petit a*, which object is not so much those things as that which causes us to want those things – our ontological lack. And if on the one hand we can see that Ruti's explication of sublimation and the Thing offers a stark contrast from the *objet petit a*, on the other, since we are back at ontological or constitutive lack, the idiosyncratic or benchmark traumas of cancer or climate grief don't compare, don't stack up – are not fungible.

Question of autotheory: my intervention

I want, too, to supplement my essentially literary-critical reading of Ruti's texts, with Anna Kornbluh's bracing critique of autotheory, part of her larger argument with respect to how the circulation of meaning matches a deprofessionalization and precaritization of the academy:

Autotheory's flexible modalities and limitless mobilities disperse it widely. In these lucrative and revered works, academics proffer lyrical expressions of personal experience and impressionistic musings punctuated by theory quips. Often shaped as aphorisms, fragments, elliptical nonnarratives, and momentary illuminations, these texts rebuff systematic elucidation. Fullnesses of charismatic persona, corporeal receptivity, and affective flooding devise an evanescent plenum that preempts criticism. The trademark theoretical posture of unassailability for the autotheorist . . . 'relies on her own vulnerability to insulate herself from scrutiny.' Immediacy in theory is this argumentless intensity immured from dissent and devoid of higher-order integration.

(Kornbluh 2024, 160)

While this full-throated critique is in some ways accurate with respect to many of the texts Kornbluh discusses, Ruti's is rife with "systematic elucidation" *and yet* also features, by dint of its origins in Ruti's own cancer diagnoses, something similar to the "theoretical posture of unassailability" – indeed, it is uncomfortably proximate to the "lean-in" feminist vulnerability that such self-help authors as Brené Brown have monetized. I think it is also possible – no, necessary – to view Ruti's turn to autotheory as itself a sublimation, an act of raising the object of her grief, to the dignity of the Thing that is the text.

It should already be obvious that this text does not obey the customary parameters of academic writing: in the pages that follow, the theoretical and the personal commingle in ways that are characteristic of contemporary autotheory. This genre interested me even before I fell ill. But my cancer diagnosis, especially the possibility that I may not live long enough to see all my active projects to completion, prompted me to write more personally, in a manner that foregrounds the affective investments that have always motivated my writing but that I, until recently, was used to hiding behind abstract rhetoric.

("BB" 131–132)

There may be some rewriting of history here: and I only say this because Ruti was in two ways doing anything but hiding behind abstract rhetoric. I am uncomfortable in aligning Ruti's queer-anti-capitalist feminism with that of Brown's liberal forms; I am trying to follow the logic of the deficiencies of autotheory that I already felt in her work, especially in books like *The Case for Falling in Love* and *Penis Envy*. But let us return to her late work, to the essays under consideration here.

Cézanne

Consider how Ruti discusses apples, first in terms of Cézanne, then the cell phone.

> Lacan was no stranger to the special delight that objects can provide. In one of my favorite sections of his seminars, he offers an example of an object that grants many people a palpable aura of the Thing: a still life of apples painted by Cézanne. Lacan explains: 'Everyone knows that there is a mystery in the way Cézanne paints apples, for the relationship to the real as it is renewed in art at that moment makes the object appear purified; it involves a renewal of its dignity.' By means of the enigmatic purification that art is capable of, apples painted by Cézanne establish a connection to the real – to the locus of jouissance – for those who view them. If there is a mystery in Cézanne's apples, it is not because he paints a perfect replica of an apple but because – by choosing to portray a slightly stylized version of an apple – he is able to capture something of the jouissance of the real in his portraiture of this mundane object. Cézanne is able to bring the viewer so close to the Thing that she is able to feel its aura: it is as if the air between the viewer and the *objet a* (as the emissary of the Thing) embedded in the painting vibrates with this aura. What makes Cézanne an extraordinary painter is that he is able to generate this vibration.
>
> ("WCNC" 12–13)

I want to juxtapose Ruti's insight – her invocation of Lacan's insight – with respect to Cézanne to, on the one hand, what the art historian T.J. Clark has to say about Cézanne's apples, and, on the other, what Ruti herself says about another object, another apple: the Apple iPhone. Actually, she only mentions the latter once, when admitting that "the object may quickly become outdated, like an iPhone that feels obsolete the minute a newer version is released" ("WCNC" 10).

In a review of a Cézanne exhibition at the Courtauld Gallery in London in 2010, Clark remarked that "Cézanne, whose work was the touchstone for critical thinking and writing on art for more than a century, cannot be written about any more" (Clark, "At the Courtauld").[9] He went on in the same article to refer to what he calls the "Cézannoia" of contemporary art world denizens as they seek in the painter a bulwark against modern life (including, presumably, smartphones) but implicitly includes himself in that group:

> Oh, but the pictures survive it all. There is an astonishing trio of large portraits of a middle-aged man, probably a farm labourer, in his Sunday best – loans from Mannheim, St Petersburg and Moscow – that it is worth travelling the earth to see hung in a row.

In Clark's formulation, on the one hand, Cézanne's apples do not, unlike the iPhone, feel obsolete[10]; on the other hand, since "the pictures survive it all," they are "worth travelling the earth" – they are worth the planet-destroying airline flights of moneyed art tourists (to which group Clark and indeed myself belong).

And, of course, the apple qua object is an overdetermined signifier or metaphor in Western metaphysics, not only for its biblical resonances but also insofar as its very polysemy triggers a crisis of meaning in the digital, and more specifically, how it works as a signifier for knowledge. Here we might juxtapose two forms of knowledge in Christianity: the "forbidden fruit" denied to Adam and Eve in Eden versus the unconscious dimension of the crucifixion that Žižek brings with his title *For They Know Not What They Do*. Deep in the forests of Wikipedia there is a debate over whether the entry for "Apple" should direct one to a discussion of the fruit or the computer company.[11] This is an example of Wikipedia's "disambiguation" protocol, whereby an article title with the potential for more than one meaning will direct the user to, in this example, the fruit or the computer company. Disambiguation thus functions as the fantasy of metalanguage, of there being an other of the Other, an Archimedean point from which one can decide on a signifier's meaning, cross the *barre* or deny castration. Ruti's apples then combine that of Cézanne, the Apple phone as lathouse or Lacanian gadget, and knowledge as poison. The sublimated object (the object raised to the dignity of a Thing, in Lacan's formulation) is not then unproblematic. Which is to say in various ways, we have the apple as the object and the Thing, or the three objects. First the *objet a* in Cézanne via Lacan as well as T.J. Clark. Then apple as knowledge that is also poison, the signifier of the lack in the other (in Lacan's algorithm, S[Ⱥ]). And then the Apple phone, the lathouse or Φ.

Hitchcock

This will seem rather compressed and jargon-y, but patience, please. The question I am asking here is whether Mari Ruti's conception of sublimation ends up functioning as a kind of denialism, a repression of the trauma or grief. To further explore this, I want to perform a kind of disambiguation on that figure in Ruti's text, that of the rings that she associates with the possibility of surviving her cancer, the rings that she purchases and then loses and then finds again. The rings that, she tells us, "carry a poignant aura of the Thing" ("WCNC" 12). The signifier "ring" has different connotations already: she later remarks that creativity qua sublimation "generates distance" between her subjectivity and her suffering, but that "[t]his distance collapses every time the bubble of creativity bursts: the doorbell rings, the phone rings, the world makes its 'ringy' demands" ("WCNC" 25). There are, then, at least two paradoxes or lapses in the "rings" that pass unnoticed in Ruti's texts: the two origin stories, and then the very real difference between rings as that which sublimates her grief, and rings that intensify that grief in their "ringy" demands.

Let us make another "ringy" demand on Ruti's theory with the figure of the tree ring. Consider a scene in Hitchcock's *Vertigo*, where Scottie and Madeleine (James Stewart and Kim Novak) encounter a giant redwood tree. Scottie tells Madeleine, "Their true name is *Sequoia Sempervirens*: always green, ever-living." "I don't like them," she says, "knowing I have to die" (Coppel and Taylor 633–664). They walk away from that tree and approach a display panel. Scottie, like a museum docent or park ranger, tells Madeleine: "there's a cross section of one of the old trees that has been cut down." Historical events are laid out on the cross section, and Madeleine touches the tree with her gloved finger – "Somewhere in here I was born, and there I died. It was only a moment for you, you took no notice."

Then she walks away, and it as if she disappears behind a tree, out of the frame of the shot. She has left the *mise en scène*, the world of the film. Keep in mind that "Madeleine" is actually Judy Barton playing the role of Madeleine, in a meta-cinematic ruse orchestrated by Scottie's friend Gavin Elster. In leaving the shot, Novak/Judy/Madeleine abandons the film spectator to the dumb Thing of nature, of the giant tree. But this is momentary; Scottie finds her, asks where she is, keeps pressing her, and finally, she says, "Take me away from here, somewhere in the light."

The forest is dark, nature is a darkness, it is not the space of cinema, of the human, it resists. But the history or chronology crudely labeled on the tree's cross section also resists, is a disregard for nature's own indifference to our dates and play-battles and settler heroes. (The Battle of Hastings, Columbus's "discovery" of America, and the 1776 Declaration of Independence are all labeled.) The tree's cross section, labeled for what Lacan calls the university discourse, is akin to a still from a film.

The cross section can also be thought of in terms of signifier of lack in the Other, the barred Other S(A̶), that I discussed earlier in terms of knowledge. Richard Boothby stresses the dualistic role, for Lacan's appropriation of Freud's *das Ding*, as "the primordial pivot around which the effects of the unconscious revolve . . . the most obscure core of the unconscious, the most inaccessible yet determinative engine of human desire" but also identified with "the most elemental functions of speech and language" (Boothby 166), reminding us that Lacan "insists, in fact, that 'the Thing only presents itself to the extent that it becomes word'" and that

> it is the way in which something of *das Ding* is taken up into and sustained by the signifier that allows for the fact, as Lacan puts it, that 'the question of *das Ding* is still attached to whatever is open, lacking, or gaping at the center of our desire'
>
> (Boothby 167)

In this reading (and also of T.J. Clark's suggestion of a taxonomy for Cézanne), the labels on the tree are not only constitutive ("the Thing only presents itself to the extent that it becomes word" [Boothby 167]) but also indicate the lack in the Thing that is also, via Žižek's "thing from inner space," our own lack.

Lacan's articulation bears an uncanny resemblance to the cross section when he remarks, in *Seminar VII*, that "the magic circle that separates us from [*das Ding*] is imposed by our relation to the signifier" (Lacan, *Seminar VII* 134). Recall the label on the cross section: "The white rings indicate the width of the tree when the various events took place." The "white rings," Ruti's rings, are themselves *das Ding*, and also, as Boothby argues, both defense from *das Ding* and our interrogative access to it. Nature as Thing is also the maternal Thing: Paul Kingsbury and I float this proposition in our introduction to *Lacan and the Environment*. "Refusing the notion that the child forsakes the mother under the father's threat of castration, Lacan proposes that the child takes its own distance from the maternal Thing, a defensive operation made possible by reliance on the signifier" (Boothby 169). Climate anxiety is not so much anxiety about the destruction of nature, but anxiety about nature as overwhelming Other, anxiety about its capacity to absorb our shit: "The most anguishing thing for the infant is precisely . . . when the mother is on his back all the while, and especially when she's wiping his backside" (Lacan, *Seminar X* 53–54, qtd. Boothby 167).

Thus far, I am reading the scene as a depiction of the human relationship with nature as nonhuman, cruelly indifferent to our petty concerns. Much the same could be said of Mari Ruti's rings, no? Madeleine sees that indifference – or, rather, Judy plays a confused woman as somehow aware of such alienation; Scottie, haranguing her like an inept psychoanalyst, nonetheless senses a *pas-tout* in her alienation. The indifference of nature to our concerns is what nature lacks: it lacks empathy, it does not care. Nature wins at the "IDGAF sweepstakes." This is to push at the possibility that Ruti holds out for her rings as sublimation and is also hinted at in an interlocutor of Lacan's in *Seminar VII*, "M. X," who points out: "The formula for sublimation that you have given us is to raise the object to the dignity of the Thing. This Thing doesn't exist to start with, because sublimation is going to bring us to it." Lacan replies to this that he is "on the right track" (Lacan, *Seminar VII* 134). Madeleine, suddenly aware of her death like Barbie at the beginning of Greta Gerwig's film, indeed like Mari Ruti when she was writing her final lectures, thus thinks of nature not as life-giving but death-dealing. That is, we can think in this way of nature as "between two deaths."

But Madeleine has been a stand-in for Mari Ruti in the past few pages: what of Ruti's position, what is the role of the Thing, of sublimation? We come back to trees and rings later – with *The Overstory* and its narrative of "the war in the woods" and with so-called Climate Gate, or the scandal over how scientists did or did not manufacture tree ring data. The tree, like a ring, is a sublime object, it is not simply nature, but in us more than us. That is, this "thing from inner space" is a direct contradiction with the idea of nature being indifferent to us. Neither position is adequate: rather, truth lies in that antagonism, or perhaps in the coincidence of opposites. Ruti's rings and her apples, Cézanne's apples, they confront us not simply with a sublimation of our grief or trauma but rather with the lack that is in nature as much as it is in us.

Notes

1 Alberto Toscano, personal communication, May 2023.

2 jebemkodyodrana. "Zizekians/Lacanians: What do you think about Mari Ruti's The Singularity of Being?" *Reddit*, 15 March 2018, https://www.reddit.com/r/zizek/comments/84qceo/zizekianslacanians_what_do_you_think_about_mari/

3 Just to give one example: in a Kleinian discussion of anxiety as a barrier against acknowledging climate change, Weintrobe describes the context for mourning loss as sometimes "a climate of hatred, bitter recrimination and relentlessness" (Weintrobe "Difficult problem," 37), rendering a metaphor for the fungibility of personal relationships versus the social (that is to say, family and friends in terms of a partner loss versus community, nation, the global for the climate) that, tin-eared, repeats the keyword of the terms of her analysis. The full title of her essay is "The Difficult Problem of Anxiety in Thinking about Climate Change" and that of the book it is in (edited by Wientrobe), *Engaging with Climate Change: Psychoanalytic and Interdisciplinary Perspectives*. This inattention to the signifier is matched by a fairly humanist and wishful-thinking conception of grief – that is, an ontologization caught up in an ego-psychology dynamic of mirroring: "[i]t . . . helps to grieve if we feel the support of those who appreciate how painful grief feels," she tells us, and "if we are able to work through our grief, with support, to the point of being able to think rationally, we have survived in heart and mind" (37). Grief here, posited as something one works through, is an object or substance wholly seen as external to the subject.

4 Indeed, Ruti seems to argue the opposite herself in *The Singularity of Being*: "the more a given subject's constitutive lack has been compounded by more specific forms of wounding, the more difficult it may be for it to claim its share of the world-making potential of the signifier" (*SB* 52).

5 Contingent or idiosyncratic lack is replaced by "benchmark trauma" in "Brokenness of Being," where idiosyncrasy is mostly used to describe desire. These distinctions are crucial to Ruti's thought and have been there since at least *A World of Fragile Things* where "ontological" (45) or "existential" lack is opposed to "circumstantial" (161n21). In *The Singularity of Being*, the distinction is between "forms of psychic energy that are constitutive of subjectivity as such (existential) and those that are circumstantial (historical or cultural)" (*SB* 47).

6 Rosemary Overell, personal communication, September 2023.

7 The poets Cecily Nicholson and Mercedes Eng have written of Emma's Acres, a community garden for "victims of crime as well

as people currently and formerly incarcerated" (Nicholson and Eng, n.p.).

8   Ruti discusses the beach and the ocean in "WCNC" only; she discusses rings and apples in both texts.

9   Clark liked this opening bit so much he returned to it in his 2022 study *If These Apples Should Fall: Cézanne and the Present* (p. 63).

10   Rather, as the title *If These Apples Should Fall* suggests, such obsolescence would be a catastrophe – the title comes from Ernst Bloch.

11   https://en.wikipedia.org/wiki/Talk:Apple. 12 September 2022. Accessed 25 September 2023.

References

Alkebulan, Paul. *Survival Pending the Revolution: The History of the Black Panther Party*. Tuscaloosa: U of Alabama P, 2012.

Boothby, Richard. "Lacan's Thing with Hegel." *Continental Thought & Theory* 2.4 (2019): 164–179.

Butler, Judith. *Subjects of Desire: Hegelian Reflections in Twentieth Century France*. 1987. New York: Columbia UP, 2012.

Clark, T.J. "At the Courtauld: Symptoms of Cézannoia." *London Review of Books*, 2 December 2010. https://www.lrb.co.uk/the-paper/v32/n23/t.j.-clark/at-the-courtauld. Accessed 24 July 2023.

Clark, T.J. *If These Apples Should Fall: Cézanne and the Present*. London: Thames and Hudson, 2022.

Coppel, Alec and Samuel Taylor. *Vertigo: Screenplay*. https://assets.scriptslug.com/live/pdf/scripts/vertigo-1985.pdf. Accessed 23 August 2023.

Edelman, Lee. *Bad Education: Why Queer Theory Teaches Us Nothing*. Durham: Duke UP, 2022.

Edelman, Lee, Alicia Arenas and Azeen Khan. "Psychoanalysis and Urgency: An Interview with Lee Edelman and Alicia Arenas." *The Lacanian Review* 6 (2018): 46–60.

Ghosh, Amitav. *The Great Derangement: Climate Change and the Unthinkable*. Chicago: U of Chicago P, 2017.

Jardine, Alice. *At the Risk of Thinking: An Intellectual Biography of Julia Kristeva*. Ed. Mari Ruti. New York: Bloomsbury, 2021.

Kornbluh, Anna. *Immediacy*. London: Verso, 2024.

Krzych, Scott. "Circumstantial Sublimation and Steven Soderbergh's *Ordinary Objects*." *Psychoanalysis, Culture & Society* 23.2 (2017): 123–140.

Lacan, Jacques. *The Seminar of Jacques Lacan: Seminar VII, 1959–60, the Ethics of Psychoanalysis*. 1986. Trans. Dennis Porter. New York: Norton, 1997.

Lacan, Jacques. *The Seminar of Jacques Lacan: Seminar X, Anxiety*. Trans. A.R. Price. Cambridge: Polity, 2014.

Lertzman, Renee. *Environmental Melancholia: Psychoanalytic Dimensions of Engagement*. New York: Routledge, 2015.

Neill, Calum. *Jacques Lacan: The Basics*. London: Routledge, 2023.

Nicholson, Cecily and Mercedes Eng. "Restorative Practices." *Black-Flash*, 22 October 2021. https://blackflash.ca/2021/10/22/restorative-practices/.

Orange, Donna M. *Climate Crisis, Psychoanalysis, and Radical Ethics*. London: Routledge, 2017.

Ruti, Mari. "The Brokenness of Being: Lacanian Theory and Benchmark Traumas." *Angelaki* 28.6 (November 2023): 123–170.

Ruti, Mari. *The Call of Character: Living a Life Worth Living*. New York: Columbia UP, 2013.

Ruti, Mari. *The Case for Falling in Love: Why We Can't Master the Madness of Love – and Why That's the Best Part*. Naperville, IL: Sourcebooks, 2011.

Ruti, Mari. *Distillations: Theory, Ethics, Affect*. New York: Bloomsbury, 2018.

Ruti, Mari. *The Ethics of Opting Out: Queer Theory's Defiant Subjects*. New York: Columbia UP, 2017.

Ruti, Mari. "The Fall of Fantasies: A Lacanian Reading of Lack." *Journal of the American Psychoanalytic Association* 56.2 (2008): 483–508.

Ruti, Mari. *Penis Envy and Other Bad Feelings*. New York: Columbia UP, 2021.

Ruti, Mari. *The Singularity of Being: Lacan and the Mortal Within*. New York: Fordham UP, 2012.

Ruti, Mari. *The Summons of Love*. New York: Columbia UP, 2011.

Ruti, Mari. "When the Cure Is that There Is No Cure: Melancholia, Mourning, Creativity." *Meaningless Suffering: Traumatic Marginalisation and Ethical Responsibility*. Eds. David Goodman and Mookie Manalili. New York: Routledge, 2024. 4–28.

Ruti, Mari. *A World of Fragile Things: Psychoanalysis and the Art of Living*. Albany: SUNY P, 2009.

Ruti, Mari and Amy Allen. *Critical Theory Between Klein and Lacan: A Dialogue*. New York: Bloomsbury, 2019.

Tom Ratekin. "Singularity, *Sinthome* and Weak Universality in Virginia Woolf's *Mrs. Dalloway* and Michael Cunningham's *The Hours*." *Singularity and Transnational Poetics*. Ed. Birgit Mara Kaiser. London: Routledge, 2015. 155–175.

Weintrobe, Sally. "The Difficult Problem of Anxiety in Thinking about Climate Change." *Engaging with Climate Change: Psychoanalytic and Interdisciplinary Perspectives*. Ed. Weintrobe. London: Routledge, 2013. 33–47.

Chapter 2

The ethics of Mari Ruti

The Ethics of Opting Out and Distillations: the asocial subject

The previous chapter concluded with a figural reading of Ruti's sublimatory rings in another medium, that of film. So too, later in this chapter, I compare fictional narrativizations of theory (the novels of Eugenides, Brand, and de Kretser) to the engagement with same we find in Ruti. This is all to make a strong argument for how cultural representations offer us comparable models for thinking not only of theoretical precepts but also for how subjects respond to the trauma of climate change. In *The Ethics of Opting Out* (*EOO*), Ruti sets forth at least three arguments: first, an exposition of "a few schematic examples" of the queer ethic of opting out (in the work of Lee Edelman, Tim Dean, Jasbir Puar, and Jack Halberstam), then, an account of the Žižekian-Lacanian ethics of the act and radical autonomy (the hegemonic example being Lacan's reading of *Antigone* in his seminar on the ethics of psychoanalysis), and, most fruitfully for what will come to be her late period work on trauma and grief, an attempt to maneuver between Edelman's "asociality" and critiques of same from within queer, BIPOC, and feminist theory. I want to spend some time on this last argument, what Ruti calls moving "beyond the antisocial-social divide" in the eponymous fourth chapter of *The Ethics of Opting Out*, but first is presented via a discussion of what connects some of the theory. As the title of her monograph indicates, Ruti is interested in queer theory's advocacy of a refusal of 21st century accommodationist moves – first articulated with respect to gay marriage. The theorists she discusses in her first chapter – from Judith Butler and Lee Edelman to Lauren Berlant and Sara Ahmed – demonstrate that "[o]pting out – the ability

DOI: 10.4324/9781003518914-3

to defeat cruel optimism, as it were – presupposes the capacity to resist what Ahmed calls the dominant 'happiness scripts' of our society, such as the marriage script" (*EOO* 18), for "the defiant subject – the subject who opts out of the system – is one who is able to turn away from the promise of happiness (as conceptualized by the normative order" (*EOO* 19). But, Ruti goes on to say, "my attitude toward the queer theoretical stance of opting out is conflicted in the sense that I see its shortcomings even as I find it conceptually engaging" (*EOO* 39). I will return to this "conflicted" stance later in this chapter.

But we can see this conflict throughout *EOO*, for even as Ruti is trying to bridge the politics of queer theory and Lacanian ethics, there is a major similarity that she neglects to point out – in part because it does not jibe with her project. Jasbir Puar's *Terrorist Assemblages* was, in the first decade of this century, a major attempt to read suicide bombers as queer subjects, not only by way of siting the body and the weapon in a kind of terrible intimacy but also for the challenges that such bodies pose to homonormative discourses of accepted queer bodies in the imperium (what, with respect to the Zionist project in Palestine, is called "pinkwashing"). Puar ranges across a variety of moments during the early 21st-century war on terror, which is to say, in the immediate aftermath of 9/11 and the U.S. state's ramping up of violent and Islamophobic militarism under the guise of self-defense, including, among other case studies, Islamophobia in British queer activism (and in *South Park*), the Abu Ghraib affair with the U.S. legalization of sodomy, misrecognition of Sikh Americans as Muslim. The terrorist assemblage is both the panoply of bodies that fall into the terrorist category for the superstate and also the specificity of what we might call the Antigone subject of the suicide bomber.

Ruti actually sounds quite liberal, if not conservative, as she attempts to narrate, but also delimit, Puar's critique of the role of queer tolerance in the U.S. imperial project:

Puar, like [Wendy] Brown, at times understates the degree to which non-Western cultures were homophobic (and misogynistic, and perhaps even racist) well before Western imperialism as well as the degree to which they remain so today, as if being "non-Western" automatically absolved a culture of all charges of bigotry; both critics tend to allow their (understandable) impulse to rescue non-Western cultures from the legacies of colonialism to overshadow their critical acumen in relation to non-Western cultural hegemonies, with the result that they, at times, make it sound like non-Western cultures are squeaky-clean – entirely devoid of the kinds of power imbalances that progressive critics have spent decades deconstructing in the Western context. One could even say that such withholding of critical acumen represents its own form of Western paternalism, whereby the non-Western "other" is either romanticized as beyond reproach or deemed to be too brittle to survive the full force of critical scrutiny. That said, I agree with Puar's problematization of the American discourse of exceptionalism, for this discourse incorrectly portrays non-Western cultures as static: forever mired in the swamp of tradition.

In the present context, the point to focus on is Puar's contention that one version of American exceptionalism is "queer exceptionalism": the conceptual alignment of American queer subjects with incomparable transgressiveness and subversiveness (precisely, with the capacity to "opt out").

(EOO 31)

What is glaring in this passage is Ruti's neglecting to actually provide any evidence for her characterization of either so-called non-Western misogyny and homophobia or Puar and Brown's "withholding of critical acumen." It's unfortunate but also symptomatic, because while Ruti proceeds to lay out a fine description of Puar's reading of the suicide bomber as constitutive critique of queer mobility and homonormative exceptionalism, Ruti is also unwilling to grant Puar's subjectless subject ("Puar reads suicide bombing as the epitome of anti-individualist politics" [*EOO* 33]) a viable trajectory. Ruti's very practice of citation from Puar, that is, flattens the text so that when Puar builds an argument from other critics, Ruti reifies Puar as author. Thus when Puar writes that "Spivak reminds us in her latest ruminations that suicide terrorism is a modality of expression and communication for the subaltern" (218), Ruti reworks the passage into *"Terrorist Assemblages . . .* suggests that suicide bombing (terrorism) 'is a modality of expression and communication for the subaltern'" (29); then, where Puar writes "Mbembe and Spivak each articulate, unintentionally, how queerness is constitutive of the suicide bomber and the tortured body" (221), Ruti claims instead "Puar . . . unhesitatingly conclude[es] that 'queerness is constitutive of the suicide bomber'" (34). This is not to disavow that Puar is in effect supporting such claims with her citation of Gayatri Chakravorty Spivak and Achille Mbembe: rather Ruti, in effect, not only substitutes a humanist subject for the asocial subject of Puar's queer terrorist but does so via the rhetoric of her (mis)reading. Further, that misreading is in the service of the very homonormative or homonationalist project of which Ruti claims to be in solidarity with Puar's critique. Here Ruti is all too clear:

> What is most relevant for our purposes is that Puar's allegiance to the Deleuzian-Guattarian ideal of the utter pulverization of subjectivity leads her to elevate the suicide bomber – whose "identity" is, literally, blown to pieces – to an icon of "queer assemblage," to assert subjects – such as the suicide bomber – who might well resent their induction to the queer nation.
>
> (*EOO* 32–33)

There are two objections. First, surely such a statement willfully misreads "identity" as the material body of the suicide bomber, rather than what Hegel calls "picture-thinking" that ascribes, on the basis of skin color, religion, sexuality, or other marker, some construct of the Lacanian imaginary. Rather, Puar wants to argue that identity qua assemblage (this is her modification of intersectionality) is constituted by the destruction of the body in the case of the suicide bomber. But it must be admitted that Puar, similarly to Ruti, misreads or misaligns identity and subjectivity: "[t]here is no entity, no identity, no queer subject or subject to queer, rather queerness coming forth at us from all directions, screaming its defiance, suggesting a move from intersectionality to assemblage" (Puar 211) insofar as identity politics qua "reactive community formations" (211) and "[i]ntersectionality demands the knowing, naming, and thus stabilizing of identity across space and time" (212). Puar's pre-emptive critique of intersectionality, notwithstanding how the latter concept has only gained traction on the left in the two decades since *Terrorist Assemblages* was first published, provides a window into what is truly troubling for Ruti.

Second, Ruti's earlier quote from Puar's citation of Mbembe and Spivak enacts a (Lacanian?) cut whereby Puar's statement that "queerness is constitutive of the suicide bomber and the tortured body" (Puar 221) becomes "queerness is constitutive of the suicide bomber" (Ruti 34). The "tortured body" is left out of Ruti's account, perhaps because it not only exemplifies a certain complexity in Puar's analysis (which, to be clear, any critical encounter must do – leave things out, but of course my contention is that the elision is symptomatic) but also because it betrays the force of Ruti's homonationalist claim that suicide bombers must needs object to being corralled into the queer nation – because, you know, Muslims are homophobes unlike Western liberals. (Please know that I am not claiming Ruti consciously subscribed to such Western fantasies. But it's there in my reading of her text.) So, there are two arguments of Puar's that Ruti neglects to cite: the tortured body, but also the possibility of the "actually existing" suicide bomber being queer. These elisions make possible Ruti's charge that Puar and Brown egregiously neglect to scold (fingerwave at) the Muslim world:

> Puar, like Brown, at times understates the degree to which non-Western cultures were homophobic (and misogynistic, and perhaps even racist) well before Western imperialism as well as the degree to which they remain so today, as if being "non-Western" automatically absolved a culture of all charges of bigotry.
>
> (*Ruti* 31)

Tellingly, this logic that one must always begin with a condemnation of the Muslim or "terrorist" or brown subject is widespread, and, indeed, discussed in Puar, when she quotes Ghassan Hage: "why is it that suicide bombing cannot be talked about without being condemned first?" (Puar 216). A similar paradox was satirized by *The Onion* in the first week of the current war in Gaza, with the headline "Dying Gazans Criticized For Not Using Last Words To Condemn Hamas"[1] – the tweeting of which article resulted in the firing of a biosciences journal editor ten days later.[2] And so we have a series of cuts enacted by Ruti's critique. Again, cutting out the tortured body but also cutting out the possibility of the "actually existing" suicide bomber being queer, all led, I argue, to not seeing the Antigone dimensions of Puar's analysis (and let me be clear that Antigone is to be found throughout not only *The Ethics of Opting Out* but Ruti's *oeuvre*). The tortured body is queer or queered in the sense, as Puar summarizes in her reading of the Abu Ghraib affair, where cultural differences (the notorious utterance "Homosexual acts are against Islamic law," coming less than a year after the lifting of the sodomy ban in the U.S. Supreme Court) were weaponized (Puar 138). Also, female suicide bombers, Puar notes, "cast out of or shunned by traditional compositions of gender and sexuality (often accused of being lesbians)" are not so much queer in identity but in their self-propelled ejection (via their exploding selves) from a stable body.

Throughout *EOO*, Ruti is working hard to argue that the various Lacanian critiques of the subject are tired, old, stale and over, rife with José Muñoz's jibe about "well-worn war chest of poststructuralist pieties," "the monotonous repetition of poststructuralist dogmas," "many poststructuralist scholars of Butler's generation seem obsessively fixated on the idea that the subject must be ground to dust" and mischaracterizations of "poststructuralist critics such as Edelman and [Lynne] Huffer [who] keep wanting to hack the subject to pieces" (87, 89, 58, 149). Ruti's error later in *EOO* as well as in her treatment of Puar is to mistake the Lacanian subject, which very much is in the tradition of the Cartesian subject, but one with an unconscious, with a caricature of the post-structuralist bogeyman. Ruti's antipathy to that critique of the subject is pertinent to this (my) book's topic, however, insofar as it is fundamental to Ruti's contradictory notions of the relationship between Lacanian lack and more contingent or socioeconomic trauma. And so, in reading Ruti's chapter "Beyond the Antisocial-Social Divide" (*EOO* 130–168), what is remarkable is her critique of Lacanian doxa that

> [T]he subject's foundational lack-in-being is the only, or even the most important, form of alienation in human life, calling attention to the myriad ways in which the subject can be injured (even 'negated') by structural inequalities such as racism, sexism, homophobia, and global economic inequalities, . . . [since for Ruti,] the recognition of the subject's constitutive lack-in-being should not . . . keep critics from acknowledging the importance of more circumstantial forms of wounding (and vice versa).
>
> (*EOO* 131)

One may argue that an ontological, a priori lack does not outweigh social inequalities but that those latter harms are simply of a different qualitative status than the constitutive, and further, that the latter are not so much negations as symptomatic of the reign of racial capital and its gendered workings. But such niceties may not convince critics who mistrust Lacanian formalism. Further, Ruti's formulation does very much anticipate, by way of its opposing logic, where Ruti's work will take her in its late stage. This is to say that by the time of "WCNC" and "BB," the social harms will themselves be relegated to a tertiary position by the contingent (in Ruti's personal case, terminal cancer). But here, a decade earlier, Ruti is both exasperated by

Edelman's intractability and – as we so often are – unwilling to see others' debates as having value or at least being significant. Ruti does, to be sure, see a connection, so we are not so much "beyond" the divide of antisocial and social theory as bridging that divide. She argues that socioeconomic trauma opens one's eyes to the constitutive, but just as I am unconvinced by her later thesis that constitutive lack better prepares us for the contingent, this logic, while reversed, similarly can be questioned. Ruti argues the following:

> Moments when something goes wrong in the concrete texture of our lifeworlds are ones when our carefully constructed fantasies collapse and we have no choice but to stare right into the abyss (in the Nietzschean sense); moments when a painful event scrambles the coordinates of everyday life force us to grapple with the fundamental uncertainties of human life.
>
> (*EOO* 132)

But I venture that it is the exact opposite: when our material or concrete existence collapses, we cling to our fantasies (of nation, identity, family) all the more strongly. Think of the reaction when people are burned out of their homes – what do they mourn, what do they wish they had rescued? They wish for their family photo albums (or, at least, before the digital: now, one grabs the laptop). There is no "traversing the fantasy," no critique of the petro-state or the Anthropocene. One remains at the bedrock of ideology: the family. Or when a disaster occurs, one sees the community suddenly rebranded: "Lytton Strong," or "Fort Mac Strong" or, most crucially, "this will not divide us": the fantasy of unity.[3] One does *not* "opt out." Nevertheless, this assertion on the part of Ruti as to the relationship of the constitutive versus the contingent (in this case, the socioeconomic) helps us better understand both the fungibility of lack and the stack (Table 2.1).

Here is a reminder of how Ruti will make the connections ten years later, in her final writings: she will say that "if we were not fundamentally lacking – if we lacked our constitutive lack – we would not have the creative, innovative capacity to endure the kinds of more contingent lacks" ("WCNC" 7), "those who have gathered an awareness of the incurable nature of their ontological lack have often also developed an array of psychic and affective strategies for coping with benchmark traumas" ("BB" 166). These phrases express much the same argument but in slightly different form, for the first does so negatively and the

Table 2.1 Ruti's theory of lack in *The Ethics of Opting Out*

1. Antisocial: ontological lack: the Lacanian psyche	Ruti: Collapse of **2** means facing **1**	In both cases, lack is fungible
2. Social: socioeconomic: race, gender, class	CB: Collapse of **2** means clinging to fantasy (denial of **1**)	

second positively. Ruti goes on to say, that is, that our very ontological lack helps us to endure (since we all lack, presumably this means that that very lack is what makes it possible for us to endure), which is slightly different from saying that those who have dealt with their lack will now be able to cope. Whereas in *EOO*, she says that it is the contingent lack or trauma that then forces us to confront our constitutive lack. The causality is reversed in the original formulation; but in all three cases, it is argued that one trauma helps us deal with another, whereas, as I argued earlier, it can be the reverse, and one trauma will so wound us that we cannot take on the other. This indeed is the argument I am making *in toto*: that the question of the fungibility of grief, of whether grief stacks, or we deal with one then another, or they make each other worse, is the most important contemporary politics of grief. We have to revise the *EOO* chart (Table 2.1). In Table 2.2, I want to offer a sense of how Ruti, and I, theorize three kinds of lacks or trauma.

First, it is important to mark those three kinds of lack, and since Ruti is arguing in "BB" and "WCNC" that the sublimation one undertakes to cope with one's Lacanian lack can also help one to deal with contingent or benchmark trauma, this new lack or trauma precedes the social, since she also argues that such sublimation or Lacanian training does not necessarily aid one in coping with social lack. So, we have this order/taxonomy:

1. ontological lack: the Lacanian psyche
2. contingent/benchmark lack: cancer, climate change
3. socioeconomic lack: race, gender, class

This is where the battle lines are drawn both privately, in our own sensorium, and in the public world. Does a forest fire in Los Angeles matter more than one in Fort McMurray? If a parent's death pulls

Table 2.2 Ruti's theory of lack in The Ethics of Opting Out and later essays

EOO		"BB"	"WCNC"	
Social: **3** matters Antisocial: **1** matters		Introduction of **2**		
Ruti: Collapse of **3** means facing **1**	CB: Collapse of **3** means clinging to fantasy (denial of **1**)	Ruti: If no **1** can't deal with **2** (doesn't help with **3**)	Ruti: Having dealt with **1** can deal with **2** (again, doesn't help with **3**)	CB: Un-unh, **1** makes things worse, in either case (**2**, **3**)

at us more than another death, what is the meaning? And this comes down to some of the key Lacanian tropes: *Che vuoi?* and do not cede your desire. To be sure, as Ruti repeatedly hammers home, "negativity is variable and unevenly distributed" (*EOO* 134), it is "unevenly distributed across the collective field" (*EOO* 170), there are "variable levels of suffering" (*EOO* 111), but this Trotskyist truism (see Warwick Research Collective) also works at the subjective level: it isn't simply that for some subjects the social or contingent lacks (racial violence, economic depredation, global neglect) are more grievous than the constitutive (which cannot be "shown," anyway, but only asserted), but that works at the level of the subject and may indeed be a way in which the subject is social. The subject is social via their grief, which is always contingent even when constitutive because of the unknowability of its prioritization or fungibility or stacking.

The subject founded on unevenly distributed trauma or lack may seem similar to the caricature of the post-structuralist anti-subject or subjectless subject, desubjectified or subjective destitution, but here three further aspects of Ruti's discussion have to be met. First is her misreading of sexual trauma or jouissance in the queer negativity of Edelman as being too literally that of sexual intercourse. Thus, after quoting Edelman on how sovereignty only occurs in moments of the "unbearable, which is to say, of our own enjoyment: the enjoyment 'we' never own" (Berlant and Edelman 9, ctd. *EOO* 138), Ruti goes on to characterize this as "the decentering jouissance of the real, the kind of jouissance that erupts in orgasmic spasms of 'enjoyment,' represents a 'traumatic encounter,' a 'shock' that renders life 'unbearable.' Unbearable? Seriously?" (*EOO* 138). But this is to willfully misread Edelman (keep in mind that "unbearable" is in the subtitle of Berlant and Edelman's book): Edelman very clearly says that for him, jouissance is as much about "shocks of discontinuity" (Berlant and Edelman 4), "sex without optimism" (15), yes, but also even a fairly scatological fantasy ("pooping back and forth" in the film *Me and You and Everyone We Know*) is both "realistically, physically . . . impossible" and something that can be conveyed with the emoticon))<>((for all of it being "a structurally impossible achievement" (23, 24). What is most telling in Ruti's pushback against Edelman here is how she appeals to what we now call one's "lived experience," indeed, her nationality or ethnicity: "My skepticism regarding this argument may be due to my Scandinavian matter-of-factness about sex, for I have never found my 'enjoyment' (jouissance) particularly unbearable" (*EOO* 138). This is hardly a convincing argument and may indeed be a textbook example of fetishistic disavowal. I know very well that Edelman is talking about unknowability, but I'm going to make a big deal out of being a straightforward Finn. Ruti's disavowal – playing the Nordic card, as it were – is also symptomatic, if I can put it that way, for as has been argued by Žižek (and also, more recently, by Steven Swarbrick), psychoanalysis actually frees sexuality from the sensual, "unhinge[s] sexuality from its object" (Swarbrick 176), for "psychoanalysis is the only discourse in which you are *allowed not to enjoy* – not forbidden to enjoy, just relieved of the pressure to do so" (Žižek 104). This is very much in line with, but also an advance on, the Slovenian thesis that sex, or the sexual non-rapport, names not the sexual act but a constitutive lack which reemerges in the dyad and is very much to do, as Edelman argues, but also Alenka Zupančič, with knowledge:

> What is at stake with this lack is thus not a missing piece of knowledge *about* the sexual (as a full entity in itself); what is at stake is that (drive) sexuality and knowledge are structured around a fundamental negativity, which unites them at the point of the unconscious. The unconscious is the concept of an inherent link between sexuality and knowledge in their very negativity.
>
> (Zupančič 142)

But to understand Ruti's argument here, it is necessary to examine how she characterizes the object of theory itself.

"Progressive critical theory"

The narrative runs as follows: first, decades ago, we believed that the resistance to theory was its own best iteration. Then, we were post-theory, post-critique.[4] With Ruti, this means her series of interventions in the 2010s – in *The Singularity of Being, The Ethics of Opting Out,* and *Distillations* – against what she calls, in the latter two volumes, "progressive critical theory."[5] That critique is both within theory, but also in some ways an exit strategy, as signaled by her later declarations in favor of autotheory. But it is also possible, following Wendy Brown's argument, to see that theory itself is a form of social sublimation. And thus Ruti will characterize her resistance to queer theory, to theory in general (one is reminded of Paul de Man's notion that the resistance to theory is inherent in theory itself), as a matter of being "levelheaded" (*EOO* 150), and therefore, again, intentionally (it seems to me), misreading Wendy Brown's assertion that theory does not describe the world as it is (a characterization quite close to Muñoz's utopianism, about which Ruti is far more charitable[6]):

> I am not sure I would go as far as to argue, as Brown does, that "theory is never 'accurate' or 'wrong'." When evolutionary psychological theory . . . claims that every aspect of human sexual behavior can be traced back to the drive to produce children, it is simply just wrong (even stupid).
>
> (*EOO* 151)

Of course, Brown is not saying all academic writing is theory: "no two disciplines or subfields mean the same thing by theory or value it in the same way," she tells us, going on to say that even in political science (the topic of the chapter in *Edgework* from which Ruti quotes: Brown 60–82), the term can range in reference "from rational choice and game theory to cultural and literary theory to analytic philosophy to hermeneutics to historical interpretations of canonical works" (Brown 80). And whereas Brown is arguing specifically for the cultural-theory influenced work of political theory, her standard is, again, less to do with whether, say, Judith Butler or Jean Laplanche (whose work she is specifically citing in the passages Ruti quotes) are accurate or wrong (or "even stupid"), but rather "it is only more or less illuminating, more or less provocative, more or less an incitement to thought, imagination, desire, possibilities for renewal" (Brown 80). Ruti, however, keeps striking the same chord, referring to "theorising that sounds like futile talk" (151), and says that "there is something meaningless about a theoretical ideal that no one – not even the author of the theory – can ever approximate" (161). This may indeed be why Ruti came to feel she was "hiding behind abstract rhetoric" (Ruti 2023, 131–132): a classic case, perhaps, of projection.

But consider how Brown continues:

> There is another reason that theory cannot be brought to the bar of truth or applicability. Insofar as theory imbues contingent or uncon- scious events, phenomena, or formations with meaning and with location in a world of theoretical meaning, theory is a sense-making enterprise of that which often makes no sense, of that which may be inchoate, unsystematized, inarticulate. It gives presence to what may have a liminal, evanescent, or ghostly existence.
>
> (Brown 81)

This sounds remarkably like the kind of sublimation Ruti calls for throughout her career, from *The World of Fragile Things* to her two late essays. Imbuing "contingent or unconscious events" – which is to say, both the constitutive and contingent traumas and lack – with mean- ing, theory as a "sense-making enterprise of that which often makes no sense" – this all seems very much like the kind of creativity as sublima- tion we have already encountered in Ruti's rings or Cézanne's apple. And, too, to bring de Man's "resistance to theory" argument (that it occurs from within theory) together with Brown's conception of theory, we can argue that Ruti's critique of "progressive critical theory" consti- tutes, both as immanent critique and exit strategy, such a sublimation. Queer theory is the trauma, the Real, that Ruti must not escape from but grapple with the bogeyman, Ruti's ex-partner if you will, that "has been relentlessly dismissive of habits, particularly of habits of thought that organize social collectivities" and with "mind-numbing regular- ity . . . undertakes the slaying of the humanist subject" (*D* 51, 54). The specificity of the tropes Ruti mobilizes here should not be overlooked, for they indicate two valences of theory: on the one hand, it is akin to one's ex who is dismissive of one's concerns, dismissive of "habits of thought that organize social collectivities" – this would be Ruti's critique of the Žižekian tendency, of Eisenstein and McGowan. On the other hand, theory conducts a murderous but also mind-numbing project that keeps "slaying" the humanist subject. In some ways, this latter charge constructs theory as a terrorist assemblage, in Puar's sense, and if Ruti wants to break up with theory (which then leads to her autotheoretical turn), she also is looking for an exit strategy in the sense of the U.S. imperial project: how to leave Vietnam, or Afghanistan, how to leave the field of battle after being defeated by the asymmetrical warfare of theory.

My conceptualizing of theory in terms of intimacy, one's partner, is similar to how the novelists narrativize cultural theory in three works of fiction from the past decade or so: in Jeffrey Eugenides's *The Marriage Plot*, in Dionne Brand's *Theory*, and in Michelle de Kretser's *Theory and Practice*. *The Marriage Plot* is the most self-consciously set texts, placed at the peak of Ivy League post-structuralism: Brown college in the 1980s. In the first third of the book, Roland Barthes's *A Lover's Discourse* vies with Victorian novel plots as the frame for a recent Brown graduate. In Brand's novel, set in the 2000s, theory is now seen as a more sinister force, and a perpetual graduate student enjoys blaming the university structure for their failure to finish indulging in such fantasies:

> I had constructed my thesis committee in the hope that they would leave me be. After all, what could these antiquated senators of the dying regime possibly give to me? Their lives had been lived in privilege and elitism. They had fooled themselves into thinking that merely because they had privilege, they had earned it. They'd never taken into account the violence their existence had perpetrated on the world, on the very people who lived around them. They'd oiled their way into schools and clubs and journals and conferences. They actually believed that this made them worthy – they confused their privilege with intellect.
>
> (Brand 150)

Theory is set in Toronto, not an American elite college town, and *Theory and Practice* similarly works at the periphery, setting off Melbourne and Sidney university communities in Australia (again, in the 1980s). De Kretser's narrator, like Brand's, is racialized (Brand's is Black, de Kretser's is South Asian) and is stricken when she realizes her thesis topic, Virginia Woolf, had retrograde – racist – attitudes toward colonial subjects. The post-structuralist "Death of the Author" seems only a way to avoid dealing with the canonical (but feminist) writer, and postcolonial theory is outside her supervisor's area of expertise ("Paula wouldn't want to supervise that," she replies. "She doesn't know the area – women's writing is her thing" [de Kretser 82]).

Ruti's theoretical argument (again, pay attention to the subtitle: that for *EOO* is "Queer theory's defiant subjects") entails a reckoning with reason and with mastery, it seems, but also one that seems to not want to talk about ideology or the unconscious. So, a defense of Western human rights rests on the very ideological-sounding, "my assumption being that people outside the Western world are not mere passive dupes of Western values, that they are perfectly capable of deciding which political goals they want to support" (*EOO* 154). Ruti continues: "I find the unqualified rejection of both reason and mastery that has become habitual in contemporary progressive theory so absurd. True, none of us is fully rational. Nor can we ever fully master anything" (155). But she is missing the point here in this blanket condemnation and generalization (which, tellingly, is not supported with any evidence). Surely we might want not so much to reject mastery but, as I tell my own students (to borrow Ruti's trope, that of referring to her own pedagogical practice), we should not try to "master" a theoretical text, be it a work of Indigenous theory, or psychoanalysis, or Afro-pessimism: not because that is morally wrong but because it assumes an unproductive form of reading, a desire for competence that, however demanded by the institution – indeed, by the big Other of one's seminar colleagues – should be resisted.[7] This resistance will of necessity take different forms. An important example in Ruti's canon, from her work of the 2010s to the final essays, is Antigone.

The act: Lacan and *Antigone* and two kinds of desire

Much of what Ruti is critical of in *The Ethics of Opting Out* is what she sees as the dead end whereby queer theory has come up against a subjectless theory that forbids any social link. This is a very patient diagnosis, but it is also one that has been with us for decades: the idea that Lacanian or Foucaultean theory is politically disabling. But this is to wish, I think, for too easy a connection between negative critique and a political program. The question of climate grief and the sociality of its fungibility may offer a way out, a way of conceiving of the negative or subjectless subject. The Lacanian subject is a subject of the unconscious, who is wrought by desires and enjoyment they do not entirely control and yet are responsible for. This is the lesson of the radical autonomy Lacan and Žižek find in the figure of Antigone, who refuses to cede her desire in the face of Creon. Lacan's reading of Sophocles asserts, first, that her ethics have to do with the image of Antigone, with her "unbearable splendor" (*Seminar VII* 247), but also with a "spectral analysis" of desire: "It is when passing through that zone that the beam of desire is both reflected and refracted till it ends up giving us that most strange and most profound of effects, which is the effect of beauty on desire" (248). This notion of beauty and its effect on desire entails not ceding her desire to the desire of the Other via the idea of going beyond a limit – what in the ancient Greek was called *Atē*, and to which Lacan devotes a great deal of time. First, he declares that Antigone's choice to perform burial rites for her brother is "what an absolute choice means, a choice that is motivated by no good" (240) in the sense that goods signifies not only a rather simplistic notion of morality but also social order, "the morality of power, of the service of goods" (*Seminar VII* 315: see also *EOO* 45–46). The beautiful, that is, is the limit of what Lacan, borrowing from Sade, calls the "second death," or the zone between life and death such as is inhabited by Antigone when she has been entombed alive. And even before that event, Antigone's entombment, we are already in a hostile environment:

Upon his return the messenger describes what happened in the following terms: first, they removed the dust that was covering the body, and then, they placed themselves up-wind so as to avoid the awful smells, because it stank. But a strong wind began to blow, and the dust started to fill the air and even, the text tells us, the heavens themselves. And at the very moment when everyone tries to escape, to cover their heads with their arms, and to go to earth at the spectacle of the change in nature, little Antigone appears at the height of the total darkness, of the cataclysmic moment. She appears once more beside the corpse, emitting moans, the text says, like a bird that has just lost its young.

(*Seminar VII* 264)

The phenomenology of dust here is striking and is repeated in quite different contexts (which resemble each other) in a 2014 interview with

Byung-Chul Han and in a 2008 novel by Xu Zechen. Discussing the globalization of production, Han remarks:

> Here in Germany, we live under an illusion. We have largely moved production elsewhere. Our computers, our clothes, our smartphones are produced in China. But the desert is coming closer and closer to Beijing and one can hardly breathe there because of the smog. When I was in Korea, I saw these yellow clouds of dust even reaching Seoul. You had to wear a protective mask – the particulates damage the lungs.
>
> (Han 132–133)

Xu's novel *Running Through Beijing* is a nice grimy novel of selling fake IDs and DVDs, hustle but also about films (including *Run Lola Run*, Chow Yun-Fat, Jackie Chan, porn). And, of course, we have the reactions to forest fires, where one can now track with one's phone the hot spots, but also has to limit one's outdoor activities hundreds of miles away from a fire because of the, as Han says, particulates in the air. Lacan tells us elsewhere that the very act of breathing involves bringing a foreign object into our body: the uncanny nature of extimacy, say. The reaction of the environment, a dust storm, in *Antigone* is redolent both of the difficulty in breathing due to climate change today and such commensurate structures as anti-Black state violence (the George Floyd protest slogan "I can't breathe") and, of course, the COVID-19 pandemic, but it also reaches back from now (or forward from Sophocles) to one of the paradigmatic cases in Freudian literature, Dora, who suffers not only from aphonia but also *tussis nervosa* (nervous cough) and dyspnea (difficulty breathing).

Han, Xu, George Floyd, and Dora are at the level of the symptom. But to return to Lacan's *Antigone*, he opposes Creon's phallic ontology, where "everything is political or, in other words, a question of interest" (*Seminar VII* 268) to a dialectics of the beautiful and the sublime that "allows the image of Antigone to rise up as an image of passion" (273). This sublime is not merely a more developed beauty but an anamorphism which includes "something decomposed and disgusting" that means Antigone does not fear (she "feels neither fear nor pity" [*Seminar VII* 258, see also 273]), indeed may be the cause of, the very dust storm that attends her violation of the law. For while Creon attends to the laws of gods and the *polis* (the latter may be tautological?), Antigone, Lacan declares, is concerned with chthonic laws, of

the earth. Here we can think of Lévi-Strauss's discussion of overvalu-
ing/undervaluing family relations – Antigone's burying Polynices is an
example of the first, Eteocles killing Polynices of the second alongside
the proposal or denial of humanity's autochthonic origins.[8] Regarding
the latter dyad, Lacan tells us the following:

> I haven't stopped emphasizing the fact that it is for the sake of her
> brother who has descended into the subterranean world that she
> opposes κήρυγμα [*kérygma* or preaching], that she resists Creon's
> order; it is in the name of the most radically chthonian of relations
> that are blood relations.
>
> (*Seminar VII* 277)

The Lévi-Straussian formula (A:B::C:D) might be extended to Creon's
son Hemon's death, which is Creon's fault, "his own mistake" (*Seminar VII* 277), but Antigone has gone beyond the *Atē*, beyond the limit,
which concerns the big Other, she has gone beyond that in not ceding
her desire: "Antigone perpetuates, eternalizes, immortalizes that *Atē*"
(282–283).

But this is the point at which, as well, we can point to a productive
contradiction in Lacan's theory of desire. As Ruti brilliantly discusses
in chapter 3 of *Distillations*, in the section "Two Types of Desire"
(*D* 105–108), a common mistake is "to approach desire from the perspective of ideology critique – which defines desire simply as an extension of dominant ideology" (*D* 105). Rather, she continues, "Lacan
operates with . . . two different definitions of desire" (105). The more
commonly known is encapsulated in the phrase, "desire is the desire of
the Other." This is worked out in Lacan's "Dialectics of Desire" *écrit*
and via the idea of extimacy. In this way, Lacan renders Freud's insights
into a more social register: so too when he argues the unconscious is
the discourse of the Other. The social solidarity here is notable and is
one of the ways in which psychoanalysis is not a psychology: it is not
simply a matter of the clinical treatment of the individual (although
it is that), but a social theory. We can think of "desire is the desire of
the Other" in different valences: the digital, the pedagogical, and the
political. When one goes onto a streaming app, say Spotify, one has the
option of choosing a musician's "radio," which is to say other musi-
cians or songs that people who listen to the first musician also listen to.
But this is all determined by the "black box" of the algorithm: the big
Other. Also, in the university literature classroom, there is something

known as a "teachable text," or one that "lands" particularly well with, say, undergraduates. In this case, the students in the class are the big Other. Then, we can think of activists organizing workers at Amazon or Starbucks or in a local social housing co-op. Here the organizer has to appeal to workers, to listen to their concerns, to identify how union certification will help them collectively address those issues (Ling and Malone). Here, the (unorganized) workers are the big Other. And in all of these cases, desire is the desire of the Other: the Spotify user's playlist, the professor's syllabus, the organizer's plan of action.

But then, especially in *Seminar VII*, Lacan's teaching starts to realize the problems if, as Ruti states, "it is extremely difficult to differentiate our desire from the desire of the Other" (*D* 106). Here he follows the logic of Antigone's decision to bury her brother, regardless of its consequences, to follow, that is, her own desire. And here Lacan founds his ethics of desire: in not giving ground relative to your desire, not ceding your desire. Although what is interesting about this proposition – as uncontroversial as it would be at any gathering of card-carrying Lacanians, who would all, ironically, have the same desire – is that Lacan makes this formulation quite late in the game, in the final lesson of *Seminar VII* (6 July 1960), and with nary a reference to Antigone. But her act is certainly read as hewing out her desire from that of Creon's, from that of the state. Be that as it may, what is perhaps most radical here is not simply that singularity of one's desire but the contradiction between these two forms of desire. An important caveat must be introduced here with respect to the radicality of Antigone's act, and this is to argue against the mainstream of Lacanian debates: I do not think Antigone's act is necessarily a suicidal one in the strict sense of the word.

It is worth noting that Ruti actually, in the line previously quoted, says that "Lacan operates with (at least) two different definitions of desire" (*D* 105). Another way to think about this, and here Lacan follows the Freud of the *Traumdeutung*, is the reflexivity of desire, the "essential dimension of desire – it is always desire in the second degree, desire of desire" (*Seminar VII* 14). This goes back to Freud's account of patients who would have dreams with the express purpose of stumping their analyst, the most famous of which is the dream of the butcher's wife. Such a notion of desire is not exactly different from that of desire being the desire of the Other (and Lacan introduces "desire in the second degree" in the same session as his discussion of the desire of the Other, the lesson of 18 November 1959). But it does muddy the waters, the distinction between the desire of the Other and not giving ground relative to your desire. We can see these difficult ethics in contemporary environmental politics in North America. Often pipelines in Canada and the United States cross Indigenous lands, whether through actual reserves, or through "traditional" or "unceded" territories. These latter terms are used more commonly in Western Canada, and in particular in the province of British Columbia, where many First Nations have never signed land use treaties with the Crown. Often in such cases, negotiated agreements will have been struck between resource companies and the elected band councils for a pipeline or other oil and gas infrastructure. But governance is complicated, and elected band councils are often viewed as colonial encumberances, and hereditary leaderships also exist, some of whom are not in favor of such agreements. And so non-Indigenous activists have to decide whether to align their actions with the democratically (however imperfect) elected band councils or the traditional (but by right of blood) hereditary leaders. Of course, there are more baleful ways in which to view this attempt at solidarity: "I shed my obdurate singularity, and in my idiocy merged with the objective world" (Comay 124). Is the solidarity of "the desire of the Other" but a way of fleeing one's own desire?

The reflexivity of desire, "the desire to desire" (*Seminar VII* 309), then, evidently has different implications for an ethics of climate activism as well as the question of climate grief: is the latter a matter of not having realized one's desire, akin to Lacan's claim that "In the last analysis, what a subject really feels guilty about when he manifests guilt at bottom always has to do with . . . the extent to which he has given ground relative to his desire" (319)? How are guilt and grief related here? Is climate grief a way of fending off guilt? I will come back to this question with respect to Žižek's adaptation of Kübler-Ross, in Chapter 4.

In addition to these questions of desire, the excursion into Lacan's *Antigone* is for two further purposes: first, to explore whether the primitive version of interpassivity that Lacan experiments here can help us understand my spectral analysis of Ruti's arguments (in *The Ethics of Opting Out* as well as her two final essays) with respect to different ontologies of trauma, and then, whether the antisocial subject of negativity that Ruti provides an overview of in *The Ethics of Opting Out* can help us further understand the ethical subject of climate grief.

Catharsis and traversing the fantasy

Throughout his discussion of *Antigone* – and elsewhere in the seminar – Lacan is concerned with Aristotle's theory of catharsis, the telos of tragedy, developed with respect to the neighbor, where he argues that the "natural basis of pity . . . depends on the image of the other as our fellow man, on the similarity we have to our ego and to everything that situates us in the imaginary register" (*Seminar VII* 196).[9] This is not far off, you will note, from his conception of Antigone's appealing as an image, but the emotional connection is mediated in a famously different direction in Lacan's discussion of how the Chorus functions:

> Next then in a tragedy, there is a Chorus. And what is a Chorus? You will be told that it's you yourselves. Or perhaps that it isn't you. But that's not the point. Means are involved here, emotional means. In my view, the Chorus is people who are moved.
>
>   Therefore, look closely before telling yourself that emotions are engaged in this purification. They are engaged, along with others, when at the end they have to be pacified by some artifice or other. But that doesn't mean to say that they are directly engaged. On the one hand, they no doubt are, and you are there in the form of a material to be made use of; on the other hand, that material is also completely indifferent. When you go to the theater in the evening, you are preoccupied by the affairs of the day, by the pen that you lost, by the check that you will have to sign the next day. You shouldn't give yourselves too much credit. Your emotions are taken charge of by the healthy order displayed on the stage. The Chorus takes care of them. The emotional commentary is done for you. The greatest chance for the survival of classical tragedy depends on that. The emotional commentary is done for you. It is just sufficiently silly; it is also not without firmness; it is more or less human.
>
>   Therefore, you don't have to worry; even if you don't feel anything, the Chorus will feel in your stead. Why after all can one not imagine that the effect on you may be achieved, at least a small dose of it, even if you didn't tremble that much? To be honest, I'm not sure if the spectator ever trembles that much. I am, however, sure that he is fascinated by the image of Antigone.
>
> (*Seminar VII* 252)

This is a well-known passage in part because of how it functions, for Žižek, as the frame on which he drapes not only his theory of interpassivity (so a sitcom's laugh track is the modern version of the Chorus: it laughs in our stead) but connects as well to the *Sujet supposé savoir* or the subject supposed to know. The reader will also note the anachronism of Lacan's little scene – this is very much a modern audience member, worried about office details, and not some fetishized, historically accurate Athenian. (He also makes a reference elsewhere in the seminar to Jean Anouilh's wartime version of *Antigone*, which some readers have mistakenly taken to indicate he believes the character to be a fascist.) And the substitution of an audience member who is "fascinated by the image of Antigone" for one who trembles indicates the imaginary nature of Lacanian emotions and affect. But we can also consider the body-blow Lacan delivers to catharsis (or purgation) alongside a more venerated psychoanalytic process, "traversing the fantasy," and Ruti's various arguments that contingent trauma means we have to "stare into the abyss" of our ontological lack (*EOO* 132), or the reverse, that our fundamental lack gifts us the "capacity to endure the kinds of more contingent lacks" ("WCNC" 7), through the very knowledge of "the incurable nature" of that lack ("BB" 166). The very motility of Ruti's thinking here – its sophistication, I would argue, its flexibility over the years – indicates, paradoxically, why it is unconvincing. It's as if she is trying different ways to argue the point but cuts the rug out from under herself. In effect, Ruti's argument is a matter of kettle logic (more on this Freudian figure later) or a snake eating its own tail.

If catharsis is akin to traversing the fantasy, things don't get better. The mistake commentators make is to think that traversing the fantasy, or the end of analysis, means some kind of improvement. If only that was the case. Remember: ordinary unhappiness is all you get. Not for nothing does Lacan stay away from actually defining this traversal. Late in the day of two texts – the *Four Fundamentals* (*Seminar XI*) and "Joyce: the Symptom" (*Seminar XXIII*) – he gets closest. Critics fall on two sides of the divide. Traversing the fantasy means "an acceptance of the fact that there is no secret treasure in me, that the support of me (the subject) is purely fantasmic" (Žižek, *Plague* 10), but the subject also "recognizes that there is no object beyond the fantasy that it might obtain . . . accepting that satisfaction derives from the fantasy itself" (McGowan 54). Traversing may simply involve, in an Edelmanesque fashion, acquiring a new *sinthome* (Coffman 55–96), but it also, Rick Boothby argues,

> does not mean that the subject somehow abandons its involvement with fanciful caprices and accommodates itself to a pragmatic

"reality," but precisely the opposite: the subject is submitted to that effect of the symbolic lack that reveals the limit of everyday reality. To traverse the phantasy in the Lacanian sense is to be more profoundly claimed by the phantasy than ever, in the sense of being brought into an ever more intimate relation with the real core of the phantasy that transcends imaging.

(*Freud as Philosopher* 275–276)

So – what? Does traversing the fantasy mean seeing through the illusions or realizing the illusions are all there is? Does it mean the end of analysis or Lacan's version of analysis terminable and interminable? These versions of traversing the fantasy, like (my reading of) Ruti's articulations of the fungibility of trauma, line up with the various "Vers" that Žižek pulled out of Freud (and which Rebecca Comay found anticipated in Hegel's *Phenomenology*) and that have been recently recalibrated by Alenka Zupančič: my argument now will be that they help us determine precisely how the subject of climate grief can also be an ethical subject. And to keep Ruti in view, her three positions also entail different valences or antinomies of lack or grief: in *EOO* she is comparing fundamental, ontological lack, psychoanalytic lack that hews dangerously close for her "softer version of posthumanist theory" (*EOO* 128) to the reckless demolition of the subject, she is comparing that to the "real injuries" of gender, race, and sexuality in particular; whereas in "WCNC" and "BB", there are three kinds of lack, or trauma, or grief lined up: the ontological, the contingent (her cancer diagnosis), and the socioeconomic.

I am arguing that those lacks are fungible, that we need to always assess how to prioritize them, how they stack (and further that how or whether we do that is in many ways unconscious or ideological: we do not necessarily "choose" or "know" that we choose or "choose to choose"[10]), and if, as Ruti argues, they necessarily subvent one another. As I said, the very differences between Ruti's formulations – her flip-flopping – suggest that fungibility (the ontological differences, a spectral analysis) and also a containerization or possibility of a silo, a firewall, as well as a deleterious, rather than supportive, intersection (or, as I've put it elsewhere, clusterfuck). This last, "negative" analysis is supported, as well, by the varieties of readings of Lacan's traversing the fantasy, not only because the literature shows significant differences, but because the substance of these traversals, or the idea that at the end of analysis one is or is not able to see one's life more clearly (to put it into self-help lingo, a vernacular

Ruti indulged/tarried in) bears not a little family resemblance to a traumatic event and its effects.

Ruti's *sinthome*

As a way of thinking of these different lacks, I now turn to Ruti's theory of the *sinthome*, a figure Lacan developed in *Seminar XXIII* when discussing the work of James Joyce. In Patricia Gherovici's *Please Select Your Gender*, she draws on the *sinthome* in her debate with the (then) dominant Lacanian reading of trans subjectivities as necessarily "psychotic." That reading went as follows: for the trans person, being "certain" that one's sexual or gender identity is of the "opposite sex," does not recognize the "big Other" of our polymorphously perverse nature of sexual identity. Thus, in a neat Lacanian patriarchy move (which of course says it is not: Catherine Millot is the critic Gherovici engages with), trans people are diagnosed as not deviant enough. Gherovici's retort is that we must depathologize trans sexualities or, more bluntly, as the title of her 2011 article declares, "Psychoanalysis needs a sex change." Her argument draws on three aspects of Lacan's theory of the *sinthome*: it "is not a complement but a supplement, it is a vehicle for creative unbalance, capable of disrupting the symmetry. The *sinthome* is what helps one tolerate the absence of the sexual relation/proportion" (*Please Select* 12). Sheila L. Cavanagh, who engages with Gherovici's work, argues that the materiality of trans gender – the surgery –means the possibility of thinking with the *sinthome*, that "substance that is subject" as Žižek puts it. But there is another *sinthome* as well. My thinking here is to try to first work out what Lacan says (*sinthome* not just as Joyce's creativity/edge of the Real but also its relation to the symptom, the Borromean knot linking the Imaginary/Symbolic/Real – I will illustrate the knot later), and then two diverging streams – Žižek and also Edelman with a kind of wild, untamed subjectivity (*sinthome* as the kernel of enjoyment upon which ideology depends), and Ruti and Gherovici (*sinthome* as creativity, as an encounter with the real that does not lead to divine violence or subjective destitution).

Let us, then, review Lacan's introduction of the *sinthome* in his twenty-third seminar, of 1975–1976. The two key aspects of that seminar are on the one hand its focus on Joyce and on the other hand the continuing development, in the 1970s seminars (the "late Lacan"), of knot theory as a way to think about subjectivity. Lacan's Joycean focus here is notable for its quasi-biographical – which is to say Romantic – account of Joyce with respect to his father – who lacks, which is why he is Joyce's *sinthome*.[11] But Lacan also – in a Joycean way that is a Lacanian trope – plays with language here and argues that Joyce "strips the sinthome of its *masaquinism*" (*Seminar XXIII* 6) or origins in Aquinas but also that Joyce "produced what I would call *sint'home Rule*" or even "the *sinthome roule*, the rolling *sinthome*, on rollers, that Joyce conjoins" (6). If at one point it seemed that various tendencies in queer, trans, and feminist strains of psychoanalytic thought and practice are the most vibrant and interesting to listen to or to read (Lacanian orthodoxy in the early to mid 2010s), and, ten years later, that it is the decolonial or anti-racist Lacan to which we harken (achieved here via the postcolonial reading of Joyce that not only aligns with Ireland's current status as one of the few countries in the European Union to be steadfast in its criticism of Israel's genocide in Gaza: aligned with the readings of Joyce attributed to Said, Eagleton, and Jameson in the 1980s, or even in the Sally Rooney of the past decade and, from the other direction, Rashid Khalidi[12]), it is also true that Lacan is both of these things (and perhaps neither) at the same time, in the same way I will argue with respect to Ruti's *sinthome*, considered with/without that of Lee Edelman.

This is because what at first glance appears to be biographical or even orthodox Freudian biography, turns out, via the role (or roll) of the "enjoy-meant" (or *jouis-sense*) of Lacan's language to be less deterministic. Indeed, even when Lacan uses the trope of Joyce's father as lacking (but especially developed later in the seminar via a fine-grained analysis of *Ulysses*) he again indulges in a wordplay that introduces the swerve of the Real:

> I'm saying that what forms the Borromean link has to be supposed to be tetradic – that perversion only means *version ver le père*, a version [turning] towards the father – and that all in all, the father is a symptom or a *sinthome*, as you wish. The ex-sistence of the symptom is merely *im-ply-cated* by the very position, the one that presupposes this enigmatic bond between the Imaginary, the Symbolic and the Real.

> (*Seminar XXIII* 11)

Lacan's figure here of the father as *sinthome* also is developed in terms of the function of the *sinthome*, which is to knot together or link together the links of the Imaginary, Symbolic, and Real. And so we can ask: Is the *sinthome* a social link? But the lesson of the *sinthome* is how it always troubles meaning and function with enjoyment. Thus, the pun of *jouis-sense* or enjoy-meant, first floated by Lacan when he discusses the jouissance of the lack in the Other (or that there is no big Other or no Other of the Other: *Seminar XXIII* 43).

Even while Lacan develops the twin engines of his discourse – a Joycean interpretation, and a theory of the knot – the connections are not clear. Here I am not making the usual complaint about Lacan's difficult style. Its difficulty has as much to do with the way he lets the style of writing (or talking) itself do the work of thinking, of theorizing, but also of course due to the circumstances of a seminar, as an oral presentation (not to mention the mediating/productive role played by Miller's editing of the transcripts, the various iterations of same in official print publication, and transcripts, recordings, or translations online).[13]

Take, for example, Lacan's oft-quoted assertion that the *sinthome* does not cease writing itself:

> I said in the past that this *possible is what stops being written.* [this is A.R. Price's translation of *Ce possible, j'ai dit autrefois que c'est ce qui cesse de s'écrire.* This last, *s'écrire*, the reflexive case of the verb *écrire*, is also translated as "writing itself".] Seeing you here in such large numbers I reckon there must all the same be some who have already heard my yarns, but you haven't remotely noticed, since I myself didn't, that a comma has to be included here. The *possible* is *what stops,* comma, *being written.* Or rather, *what would stop, taking the path of being written* in the event that the discourse I have mentioned would at last come to be, a discourse such that it is not semblance.
>
> (*Seminar XXIII* 5)

Lacan argues that "what stops being written" also means what stops being, existing, or because it is written, perhaps stops rolling (which in Lacan's French also mean stops home rule, a British policy in Ireland, and hence involves decolonial possibilities). He has already worked this field in *Encore* (*Seminar XX*): "stops not being written" (*cesse de ne pas s'écrire* – the contingent) versus "doesn't stop being written" (*ne cesse pas de s'écrire* – the necessary) and then "doesn't stop not being written (*ne cesse pas de ne pas s'écrire* – the impossible)."[14] In this modal matrix, the fourth position is what Lacan develops in *Seminar XXIII*, with *the possible*, which "stops being written" or "stops, being written."

For Edelman and Ruti (and, in a different way, Gherovici), the assertion is key to their use of the *sinthome*, and they draw very different conclusions. For Edelman, this ceaselessness has to do with *jouissance* and the horrible fixation of the subject: "Lacan, who will subsequently describe the *sinthome* as 'not ceasing to write itself,' implies from the outset its relation to the primary inscription of subjectivity and thus to the constitutive fixation of the subject's access to jouissance" (Edelman *No Future* 35). For Ruti, "Lacan carries out a close reading of James Joyce, postulating that the *sinthome* is a peculiar kind of signifier that 'does not cease to write itself'" (*SB* 115). Gherovici turns these ideas of writing to look at writings by transsexuals (memoirs), considering, first of all, the ways in which transsexuals consider being "read" by others, and then how they write: transsexuals' "two diverging uses of *reading* for gender presentation, one as not passing and the other as a call for interpretation, are both encompassed by Lacan's concept of the letter, writing, and nomination" and, further, Gherovici's "thesis is that a . . . tension between the demand for singular recognition and universal agency can be observed in transsexuals, more precisely when they write" (Gherovici, *Please Select* 216–217).

Here what I am interested in is the relation of this characterization of the *sinthome* as not ceasing to write itself/to be written to, on the one hand, its function as social link and, on the other, an interpretation of Joyce. This connection has relevance to Ruti's late-career valorization of creativity as a response to trauma, but there is a third role or notion of the *sinthome* in the seminar, one having very much to do with its appropriation in theorizing trans subjectivities: sexual difference, or the impossibility of the sexual relation as worked out in *Encore*:

> It is based only on the written in the sense that the sexual relationship cannot be written (*ne peut pas s'écrire*). Everything that is written stems from the fact that it will forever be impossible to write, as such, the sexual relationship.
>
> (*Seminar XX* 35)

Impossibility is part of the modalities (the possible and the impossible, the contingent and the necessary) that frame Lacan's various iterations of ceasing/not ceasing to write/not be written from Seminars XX to XXIII. The non-relation is raised in the session of 11 February 1976

(*Seminar XXIII*), where Lacan moves from discussing the knots to declare:

> At the level of *sinthome*, there is not, therefore, any equivalence to the relation between green and red, if we make do with this straight-forward designation. To the extent that there is a *sinthome*, there is no sexual equivalence, that is to say, there is relation.
>
> (*Seminar XXIII* 84)[15]

This is the place to disagree with Ruti's assertion that the *sinthome* is a "tight knot of *jouissance* that encloses the subject's fundamental fantasy" (*SB* 60). One way of reading Lacan in this matter is to see the *sinthome* as the loose loop connecting the rings of the Imaginary, Symbolic, and Real in the way popularized by various illustrations. Or, rather, we can think about sexual difference (say, the phallic versus the non-all) as a way to conceive of the two different formations of the *sinthome*: Ruti's formulation is phallic, and that in the previous quote from Lacan is the non-all. Indeed, this conception of the *sinthome*, and its relation to the question of sexual difference, surely authorizes its later appropriation by trans theorists.

Lacan further establishes the *sinthome* with respect to sexual differ-ence by arguing that the *sinthome* "is very precisely the sex to which I do not belong, that is, a woman" so "*a* woman is a *sinthome* for every man," but it does not work the other way around, "since precisely the *sinthome* is characterized precisely by non-equivalence" and, indeed, that nonequivalence "is the only thing . . . in which what is known as sexual relation in the parlêtre, in the human being, finds a support" (*Sem 23* 84).

For Lacan, then, the *sinthome* is the solution to the problem of the relation that is not a relation, the *sinthome* that is also for Joyce a form of creation that negates or obviates his psychosis and somehow ties together (an overdetermined metaphor: in the system of Borromean knot, the *sinthome* is that which ties together the rings of the Imaginary, Symbolic, and Real) Catholicism (Thomas Aquinas), anti-colonialism (*sinthome-roule*), and writing qua creativity. But Ruti's *sinthome* does more work itself: it argues against the "subjective destitution" school, relies on concepts of literary production that (like Lacan's) are in danger of fetishizing the author qua maker of meaning (unlike, say, meaning being produced in reading), and can finally be seen as leading to her late-stage work on sublimation and *das Ding* qua creativity as a response to trauma. This means to think about the *sinthome* with respect to enjoy-ment (*jouis-sense*) for its own sense but also the problem of uplifting the modernist artist-hero under the guise of the singular, and the question of whether the *sinthome* and the signifier are inherently asocial.

We can trace this path of enjoy-meant and subjective destitution via Edelman, for whom *sinthomosexuality* speaks, as neologistic signifier,

> to the "sin" that continues to attach itself to "homosexuality" . . . and materializes the threat to the subject's faith that its proper home is in meaning, a threat made Real by the homosexual's link to a less reassuring "home": the *sinthome* as site of a jouissance around and against which the subject takes shape and in which it finds its consistency.
>
> (Edelman, *No Future* 38–39)

The *sinthomosexual* is that figure, then, on which the heteronorma-tive hegemony lays its murderous gaze precisely because of how that queer figure stands in for how the subject *qua* subject relates to the absence of the sexual relation, to our subjectivity as lack. But also, as Edelman notes, our relation to meaning: we want our lives to mean something, and the *sinthome* is that trope or sign which refuses mean-ing, can only be enjoyed: enjoy-meant.

Ruti seeks to counter Žižek's and Edelman's emphasis on the *sinthome* and subjective destitution with a notion of the *sinthome* in relation to singularity (*SB* 60); also, her argument with respect to *jouis-sense* has more to do with creativity in general and not the trou-bling parasitism of Edelman. For Ruti, that is, the *sinthome* works as a way to critique both Žižek's appeals to the ethical act (subjective desti-tution or divine violence via Antigone) and Edelman's queer insistence on the presence:

> Yet if we allow for the possibility that the signifier does not invari-ably obey the dictates of the big Other, and that the unruly energies of the real can regenerate, rather than merely weaken, the symbolic, it becomes apparent that the signifier is not always an instrument of ideological interpellation.
>
> (*SB* 119)

Yet (I add), I am not sure that any Žižekian would dispute that the "un-ruly energies of the real can regenerate . . . the symbolic" – that, it seems to be, is exactly what the Real does via the *sinthome* (the nationalist's enjoyment supports the state, the fan's enjoyment supports capitalist sports, the academic's enjoyment supports the neoliberal university). And, to be sure, Ruti presents this argument by stating that "those interpreters of Lacan who ignore this component of his theory end up producing overly dispirited theories of subjectivity" (*SB* 118). If Ruti means the reading of Joyce that Lacan produces is ignored – certainly by Žižek, not so much by Edelman – it is worth noting that Ruti's analysis in *The Singularity of Being*, for all its attempts to construct a post- or anti-humanist theory of the subject (the singular), often, due to its fidelity to the difficulties of political trauma, is quite dispirited as much as Edelman or Žižek.

Edelman has answered some of Ruti's criticisms, although with respect to the concept of the future, in *Bad Education*, but we would do well to remember how the focus on the *sinthome* as creativity, first in Lacan's reading of Joyce and then its uptake in Gherovici and Cavanagh (which is contradicted by Edelman), can lead us in some unpleasant directions. In some ways, I think, this has to do with the ambivalence (or call it dialectics) in Lacan, that *virgule* (comma) he talks about. Ruti, in her final work, makes the argument for sublimation and creativity as a way to deal with trauma (she was writing about her cancer diagnosis, and I have been attempting to relate such creativity to responses to the climate crisis). But we also might think of the (excessive) populist subject (to adopt Molly Rothenberg's title: Rothenberg also sees political value in creativity and the *sinthome*). Is not the Trumpian voter as dedicated to creativity, the ISIS militant Shamsud-din Jabbar who set an out-of-office reply on his work email, do they not enjoy their *sinthome* (Bekiempis)? There are three (or perhaps four) subjects at play here: Joyce as modernist author, the trans subject, the *sinthomosexual*, and the populist subject. Can we imagine the social link that is dissolved (to use a trope beloved of Badiou and Edelman) by the *sinthome* which in turn subvents the populist enjoyment? Or is this more of a methodological conundrum: Ruti seldom quotes directly from Lacan and is given to according authority to the secondary critics with the locution "reminds us."[16] That is, a close reading of Lacan as demonstrated earlier argues, first, that the basic locution "does not stop being written" when subjected to the Lacanian cut, the virgule, the comma, changes from not stopping being written to the possible as that which stops and which is being written, *but not that which stops being written*. And contra Ruti, this symptom of the

Table 2.3 Ruti's theory of lack in *The Ethics of Opting Out* and later essays

EOO		"BB"	"WCNC"	
Social: **3** matters Antisocial: **1** matters		Introduction of **2**		
Ruti: Collapse of **3** means facing **1**	CB: Collapse of **3** means clinging to fantasy (denial of **1**)	Ruti: If no **1** can't deal with **2** (doesn't help with **3**)	Ruti: Having dealt with **1** can deal with **2** (again, doesn't help with **3**)	CB: Un-unh, **1** makes things worse, in either case (**2, 3**)

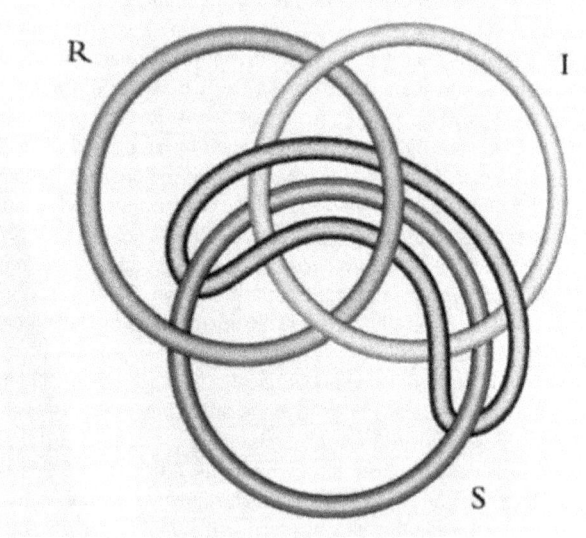

Figure 2.1 Borromean knot with *sinthome*

sinthome, this *jouis-sense* of the Real of Lacan is destructive of meaning, and of the social.

We can test the veracity of this last assertion by returning to Table 2.3, from earlier in this chapter, where I set up three lacks.

Another way to think of these lacks is with the very theory of the *sinthome*, illustrated with Figure 2.1, based on Lacan's *Seminar XXIII*.

The knot here denotes the Real, Imaginary, and Symbolic registers, with the *sinthome* indicated by the loop joining them. In this diagram, the Real signifies Lacanian lack (the psychic), the Imaginary that of socioeconomic conditions, and the Symbolic that of contingent or benchmark lack. The *sinthome*, that which holds them together, then, *à la* the discussion of creativity via Gherovici, denotes sublimation. For Ruti, sublimation would hold together the Real and the Symbolic; for my dissent, it does no such thing. We see further how such arguments hold when we turn to other cultural examples, in Chapter 3.

Notes

1 "Dying Gazans Criticized for Not Using Last Words to Condemn Hamas." *The Onion*, 13 October 2023, https://theonion.com/dying-gazans-criticized-for-not-using-last-words-to-con-1850925657/. Accessed 6 March 2025.

2 Quinn, Ryan. "Editor Fired after Sharing 'Onion' Article on Israel, Hamas." *Inside Higher Ed/Higher Education News, Events, and Jobs*. 24 October 2023, https://www.insidehighered.com/news/faculty-issues/academic-freedom/2023/10/24/editor-fired-after-sharing-onion-article-israel. Accessed 6 March 2025.

3 The meme "[Community X] strong" is evidently an example of negation or denial for it only is uttered once a disaster (forest fire, mass shooting) has taken place, once the community, that is, has had to confront its fragility. See McKinley. Whereas Ruti's argument is precisely that such a "fall of fantasies" leads to such a confrontation, that "moments when a painful event scrambles the coordinates of everyday life force us to grapple with the fundamental uncertainties of human life" (*EOO* 132) my assertion, supported by such utterances with respect to strength or not being divided, leads to the opposite conclusion: the painful event reinforces our fantasies.

4 See my discussion, with respect to "surface readings" in literary criticism, and "post-theory" in film studies, in Burnham 2016, 89–93.

5 Ruti will also, strategically, in *Distillations* distinguish between Lacanian and progressive theory on the basis of attitudes toward universalism.

6   Some queer subjects might object to Muñoz's claim that "we have never been queer," but his emphasis lies on the idea that because queerness is inherently open ended, because it cannot be fixed into a definition, it "allows us to see and feel beyond the quagmire of the present."

(*EOO* 170–171)

7   For me, the canonical statement is still Lacan's, from "The Instance of the Letter": his writing is

distinguished by a prevalence of the *text* . . . which allows for a kind of tightening up that must, to my taste, leave the reader no other way out than the way in. This, then, will not be a writing in my sense of the term.

(*Écrits* 412)

Also, of course, note Bruce Fink's masterful commentary in *Lacan to the Letter* (2004), 65.

8   Lévi-Strauss offers the algorithm A:B::C:D, or overvaluing family is to undervaluing as autochthony is to its denial in the essay "The Structural Study of Myth." In *Structural Anthropology*, 206–231; 213–217. The collection was first published in French in 1958; Lacan discusses *Antigone* in May–June 1960 and refers to Lévi-Strauss numerous times throughout the entire seminar.

9   See my "*Naked Lunch* and the Neighbor."

10 See, for this last, Alain Badiou on Kierkegaard in his *Lacan* seminar (New York: Columbia UP, 2018) sessions of 15 March and 5 April 1995; and Ronald Bogue, "To Choose to Choose – to Believe in This World," in *Afterimages of Gilles Deleuze's Film Philosophy*, ed. D. N. Rodowick (Minneapolis: University of Minnesota Press, 2010).

11 Gherovici deftly navigates the question of Lacan's reading of Joyce:

Even if Lacan tends to read Joyce's entire *oeuvre* as a memoir in a questionable biographical reading that verges on psychobiography, I argue that this approach to Joyce's work as a memoir serves above all to emphasize its function as *sinthome*.

(Please Select 232)

12 The comparison [of Palestinians] with the Irish, the only people to succeed in (partially) freeing themselves of colonial rule between World Wars I and II, is striking. In spite of divisions in their ranks, their clandestine parliament, the Dail Birann,

their nascent branches of government, and their centralized military forces ultimately out-administered and outfought the British.

(Khalidi 63)

13 I talk about these issues in "Lacan's Trash Talk: Three Objects for the Internet," in Burnham and Kingsbury.
14 Lacan, *Seminar XX*, 94. Ctd. in my "Lacan's Trash Talk: Three Objects for the Internet," in Burnham and Kingsbury, 79.
15 The green and red refer to the Imaginary (green) and Symbolic (red: the real is blue) rings or string or knots (which are reproduced in color in the French edition and the Polity translation) through which the *sinthome* wends its way.
16 Harari in *SB*; Nietzsche and Foucault in *A World of Fragile Things*; "the shrink who reminds you of your father" in *The Case for Falling in Love*; Levinas, McGowan, Eisenstein, and Žižek in *D*; Edelman and Lynne Huffer in *EOO*; Maggie Nelson in "WCNC."

References

Bekiempis, Victoria. "New Orleans Attacker Fell into Extremism after Marital and Financial Woes." *Guardian*, 4 January 2025. Accessed 4 January 2025.

Berlant, Lauren and Lee Edelman. *Sex, or the Unbearable*. Durham: Duke UP, 2013.

Boothby, Richard. *Freud as Philosopher: Metapsychology After Lacan*. New York: Routledge, 2001.

Brand, Dionne. *Theory*. Toronto: Penguin Random House, 2018.

Brown, Wendy. *Edgework: Critical Essays on Knowledge and Politics*. Princeton: Princeton UP, 2006.

Burnham, Clint. "*Naked Lunch* and the Neighbor." *Breathless Days: 1959–1960*. Eds. John O'Brian and Serge Guibault. Durham: Duke UP, 2017.

Burnham, Clint and Paul Kingsbury, Eds. *Lacan and the Environment*. London: Palgrave, 2021.

Coffman, Chris. *Queer Traversals: Psychoanalytic Queer and Trans Theories*. London: Bloomsbury Academic, 2022.

Comay, Rebecca. *Mourning Sickness: Hegel and the French Revolution*. Stanford: Stanford UP, 2010.

de Kretser, Michelle. *Theory and Practice*. New York: Catapult, 2024.

Edelman, Lee. *Bad Education: Why Queer Theory Teaches Us Nothing*. Durham: Duke UP, 2022.

Edelman, Lee. *No Future: Queer Theory and the Death Drive*. Durham: Duke UP, 2008.

Eugenides, Jeffrey. *The Marriage Plot*. New York: Fourth Estate, 2011.

Gherovici, Patricia. *Please Select Your Gender: From the Invention of Hysteria to the Democratizing of Transgenderism*. New York: Routledge, 2010.

Gherovici, Patricia. "Psychoanalysis Needs a Sex Change." *Gay & Lesbian Issues and Psychology Review* 7.1 (2011): 3–18.

Han, Byung-Chul. *Capitalism and the Death Drive*. Trans. Daniel Steuer. Cambridge: Polity, 2021.

Khalidi, Rashid. *The Hundred Years' War on Palestine*. New York: Macmillan, 2020.

Lacan, Jacques. *The Seminar of Jacques Lacan: Seminar VII, 1959–60, The Ethics of Psychoanalysis*. 1986. Trans. Dennis Porter. New York: Norton, 1997.

Lacan, Jacques. *The Seminar of Jacques Lacan: Seminar XX Encore: On Feminine Sexuality, The Limits of Love and Knowledge, 1972–73*. Trans. Bruce Fink. New York: Norton, 1999.

Lacan, Jacques. *The Seminar of Jacques Lacan: Seminar XXIII. The Sinthome*. 1975. Trans. A.R. Price. Cambridge: Polity, 2016.

Lévi-Strauss, Claude. *Structural Anthropology*. 1958. Trans. Claire Jacobsen and Brooke Grundfest Schoepf. New York: Basic, 1963.

Ling, Steff Hui Ci and Bobby Malone. "From Inquiry to Town Hall: Art Workers on Worker Identity and Political Formation." *Historical Materialism Conference*. Birkbeck: University of London, 9 November 2024.

McGowan, Todd. *Emancipation After Hegel*. New York: Columbia UP, 2019.

McKinley, Steve. "'We are Lytton, We Are Strong': B.C. Town Issues Statement Outlining Extent of Wildfire Devastation." *The Toronto Star*, 7 July 2021. https://www.thestar.com/news/canada/we-are-lytton-we-are-strong-b-c-town-issues-statement-outlining-extent-of-wildfire/article_a1392f06-b375-5671-8851-13ae7b031cb0.html. Accessed 19 May 2025.

Puar, Jasbir. *Terrorist Assemblages: Homonationalism in Queer Times*. 2007. 2nd. Ed. Durham: Duke, 2017.

Ruti, Mari. "The Brokenness of Being: Lacanian Theory and Benchmark Traumas." *Angelaki* 28.6 (November 2023): 123–170.

Ruti, Mari. *The Call of Character: Living a Life Worth Living*. New York: Columbia UP, 2013.

Ruti, Mari. *The Case for Falling in Love: Why We Can't Master the Madness of Love – and Why That's the Best Part*. Naperville, IL: Sourcebooks, 2011.

Ruti, Mari. *Distillations: Theory, Ethics, Affect*. New York: Blooms-bury, 2018.

Ruti, Mari. *The Ethics of Opting Out: Queer Theory's Defiant Subjects*. New York: Columbia UP, 2017.

Ruti, Mari. "The Fall of Fantasies: A Lacanian Reading of Lack." *Journal of the American Psychoanalytic Association* 56.2 (2008): 483–508.

Ruti, Mari. *The Singularity of Being: Lacan and the Mortal Within*. New York: Fordham UP, 2012.

Ruti, Mari. "When the Cure Is that There Is No Cure: Melancholia, Mourning, Creativity." *Meaningless Suffering: Traumatic Marginalisation and Ethical Responsibility*. Eds. David Goodman and Mookie Manalili. New York: Routledge, 2024. 4–28.

Ruti, Mari. *A World of Fragile Things: Psychoanalysis and the Art of Living*. Albany: SUNY P, 2009.

Swarbrick, Steven. *The Environmental Unconscious: Ecological Poetics from Spenser to Milton*. Minneapolis: U of Minnesota P, 2023.

Tom Ratekin, "Singularity, *Sinthome* and Weak Universality in Virginia Woolf's *Mrs. Dalloway* and Michael Cunningham's *The Hours*." *Singularity and Transnational Poetics*. Ed. Birgit Mara Kaiser. London: Routledge, 2015. 155–175.

Xu, Zechen. *Running Through Beijing*. 2008. Trans. Eric Abrahamsen. San Francisco: Two Lines, 2014.

Žižek, Slavoj. *How to Read Lacan*. 2006. New York: Norton, 2007.

Žižek, Slavoj. *The Plague of Fantasies*. 1997. London: Verso, 2008.

Zupančič, Alenka. *What Is Sex?* Cambridge: MIT P, 2017.

Chapter 3

Cli-fi and aesthetics

Cli-fi: Amitav Ghosh prolegomenon

Turning from the last chapter's intricate discussion of Ruti's theory, I want now to consider how cultural objects – novels, art – respond to the climate crisis. Here we have a replay of the problem of whether climate grief is itself a response to lack, or trauma (and, hence, an emotion that stands in for something else, that is a screen) or is that foundational, contingent, or ontological trauma itself and, further, how we can think of cultural objects as a kind of sublimation – the creativity Ruti calls for. That creativity may apply to the works of art or literature themselves: we will see how artists continue to turn detritus into art, rather like the other great example, besides Cézanne, in Lacan's *Seminar VII*: Jacques Prévert's matchbox frieze. Lacan describes visiting Prévert, the Surrealist poet:

> the match boxes appeared as follows: they were all the same and were laid out in an extremely agreeable way that involved each one being so close to the one next to it that the little drawer was slightly displaced. As a result, they were all threaded together so as to form a continuous ribbon that ran along the mantlepiece, climbed the wall, extended to the molding, and climbed down again next to a door. I don't say that it went on to infinity, but it was extremely satisfying from an ornamental point of view.
>
> (*Seminar VII* 114)

Lacan's example, in its ontological trashiness, rhymes with another well-known anecdote, that of glimpsing a sardine can floating in the ocean and suffering a crisis of not belonging (*Seminar XI* 95–96). And

DOI: 10.4324/9781003518914-4

the matchbox already is an object that has at its center a void or hole (and also, arranged around the poet's room with each box's tongue sticking into the next one, has a certain carnal dimension, a "copulatory force" as Lacan puts it [*Seminar VII* 114]). Ruti points out, in turn, that

> Lacan introduces two other examples that capture the relationship between lack and signification, namely an empty mustard pot and a hollow piece of macaroni ("a hole with something around it"). His point about these mundane "objects" is that they render concrete a fundamental fact about human life, namely that "the fashioning of the signifier and the introduction of a gap or a hole in the real is identical." This is a precise way of expressing what I have emphasized all along, namely that the lack in the real that results from the subject's encounter with the symbolic world is also what makes it a subject of signification.
>
> (*SB* 128)

But creativity also is made the subject matter of the work, as in Richard Powers's *The Overstory*, when an activist, after abandoning a violent group, turns to memoir writing as a practice. Within climate fiction, questions of the fungibility of grief are rehearsed, as I argued briefly in the introduction, with respect to Fagan's *The Sunlight Pilgrims*, where a character prefers climate disasterism to the trauma of transphobia. Here we also encounter the problem of library categories insofar as that they rely on "aboutism," on a book being categorized by its subject, not its form. The Library of Congress and American Library Association subject headings for *The Sunlight Pilgrims* list the novel in terms of "Survivalism-fiction" and "Dystopias," respectively. Now, there are a number of problems with reading literature in terms of such content. Survivalism, only a minor motif in the novel, acquires a self-fulfilling logic, and dystopia may actually be a form of utopia (Moylan 147). But the nominalism or reification of the category admits no such nuance. This category, or mediation, problem is not merely one to be laid at the feet of library practices: consider Amitav Ghosh's influential book from 2016, *The Great Derangement: Climate Change and the Unthinkable*, in which he argues that while culture – literature, art – is the cause of climate change, it cannot help us to tackle that problem:

> Culture generates desires – for vehicles and appliances, for certain kinds of gardens and dwellings – that are among the principal

drivers of the carbon economy. A speedy convertible excites us neither because of any love for metal and chrome, nor because of an abstract understanding of its engineering. It excites us because it evokes an image of a road arrowing through a pristine landscape; we think of freedom and the wind in our hair; we envision James Dean and Peter Fonda racing toward the horizon; we think also of Jack Kerouac and Vladimir Nabokov. When we see an advertisement that links a picture of a tropical island to the word paradise, the longings that are kindled in us have a chain of transmission that stretches back to Daniel Defoe and Jean-Jacques Rousseau: the flight that will transport us to the island is merely an ember in that fire. When we see a green lawn that has been watered with desalinated water, in Abu Dhabi or Southern California or some other environment where people had once been content to spend their water thriftily in nurturing a single vine or shrub, we are looking at an expression of a yearning that may have been midwifed by the novels of Jane Austen. The artifacts and commodities that are conjured up by these desires are, in a sense, at once expressions and concealments of the cultural matrix that brought them into being. This culture is, of course, intimately linked with the wider histories of imperialism and capitalism that have shaped the world. But to know this is still to know very little about the specific ways in which the matrix interacts with different modes of cultural activity: poetry, art, architecture, theater, prose fiction, and so on. Throughout history these branches of culture have responded to war, ecological calamity, and crises of many sorts: why, then, should climate change prove so peculiarly resistant to their practices?"

(Ghosh 9–10)

"What is it," Ghosh asks a bit later, "about climate change that the mention of it should lead to banishment from the preserves of serious fiction? And what does this tell us about culture writ large and its patterns of evasion?" (11).

Now, Ghosh makes any number of attempts to answer this question, but in a way the question's frame provides its own answer: that is, again, a matter of the category, of "serious fiction," by which he means literary fiction, the middlebrow, but not science fiction. His reasoning is almost tautological – since realist fiction deals with the ordinary – albeit for a certain middle-class notion of same – it cannot conceive of such extraordinary events as floods, tornadoes, or forest fires (or an unprecedented cold winter) – and therefore such events can only be accommodated in science fiction, fantasy, or horror.

We have to expand or dialectically contradict Ghosh's argument, and such a reading turns out to be one demanded by Ghosh himself. For is he not, in making his argument that "serious fiction" cannot but should deal with climate fiction, also implicitly arguing that literary criticism can and should be – indeed is, in his book – capable of doing so? Also, we can think of how Ghosh is himself repeating, in a *fort-da* moment, his own earlier argument with respect to another non-representation:

> In his prophetic 1992 essay "Petrofictions," author Amitav Ghosh famously laments the lack of fictions addressing oil and what he terms the "Oil Encounter" – the historic intertwining of the fates of Americans and the peoples of the Middle East over this resource. Ghosh offers multiple reasons why the sixteenth-century spice trade – his point of comparison – generated greater fictions . . . and more of them than has oil, including the professionalization of contemporary fiction, which he claims has to come to focus on "a stock of themes and subjects, each of which is accompanied by a well-tested pedagogic technology."
>
> (Szeman 3)

But to understand all of this, we have to think about literature qua literature, that is, again, its *non-dit*, its *impensée*, as Jameson called it, its political or environmental unconscious qua form. And yet that inquiry may also, like an oil well, run dry. Mark Bould's *Anthropocene Unconscious*, for instance, is tragically incapable of reading via formal innovation, as this take on Lucy Ellman's novel *Ducks, Newburyport* makes clear:

> It is as if cable news, tabloid journalism and talk radio – as well as some more reliable sources – stream through the narrator, weaving the centrifugal into the centripetal, capturing something of the way in which human interiority is infiltrated and shaped by exteriority.
>
> Thus despite *Ducks, Newburyport*'s inwardness, unfolding anthropogenic crises run through it: extractionism, fossil fuels, dirty power, fracking and global warming; agribusiness, factory farming, consumerism and carbon footprints; melting icecaps, extreme weather, floods, and lowlying islands threatened by rising tides; endangered species, coral die off, colony collapse, and the sixth great

extinction; the displacements, migrations and behavioural changes of various species; viral pandemics, cancer clusters, respiratory diseases; declining air quality, diminished watersheds, dams, dead lakes and habitat destruction; chemical, industrial and domestic pollution; isotopes, lead, mercury, microbeads, microplastics, nanoparticles, PCBs, radiation, tailings and toxins; population growth, slave labour, climate migration, refugees, camps and cardboard cities.

In fact, it has no intention of avoiding such things, but the stream-of-consciousness form Ellmann deploys makes constructing a coherent critical account of them impossible. Rather, it establishes their typically unacknowledged presence in and impact upon ordinary, if still relatively privileged, lives. By repetition and accretion, the Anthropocene unconscious emerges into visibility.

(Bould 43)

It may be that what Bould needs here is a Lacanian theory of extimacy ("the way in which human interiority is infiltrated and shaped by exteriority" indeed!) and to kick to the curb the normative reifying university discourse that wants to "construct a coherent critical account." But I don't want to fall into the sci-fi special pleading with which Jeff VanderMeer, for instance, greets Ghosh's *The Great Derangement*:

Ghosh appeared to have built a thesis based on, quite frankly, not reading enough fiction beforehand, and a misunderstanding of the often artificial marketing construct that is "genre" fiction. At the same time, I feel not a bit of territorialism about the idea that writers outside of speculative fiction should tackle climate change.

(VanderMeer)

I should add that I agree with VanderMeer's critique in the same article of Powers's *The Overstory*. In order to explain this argument, bringing together readings of Ghosh as well as questions of climate fiction and its relation to climate grief, it is helpful to draw some more on Jameson to think of how climate or the environment is represented via the forbidding of graven images:

1. To paraphrase Jameson, if everything stands for something else, then so too does cli-fi.
2. Or: following Jameson on Freud: all dreams are about climate change, except dreams about climate change.

3. Or: do a Blochian distinction between secular climate doomism (content) versus theological (form/narrative).

To expand these propositions briefly, whereas Jameson's original *bon mot* declared that "if everything means something else, then so does technology" (*Geopolitical Aesthetic* 11), such a reversal necessarily works both ways: fiction may not be directly about the climate crisis but still connotate such – this would be a reading of a text for its environmental unconscious; but also, what seems to be about a forest fire or tsunami or the global garbage crisis will invariably concern itself with how those concerns manifest themselves at a human scale – rather like how, when the analysand enters the clinic with their "presenting symptom," there is a lot of work to track underlying desires and anxieties. Consider Rogelio Braga's short story "Fungi," in which Filipino characters scavenge what appears to be laundry detergent from a garbage dump called the "Promised Land," thinking the soap possesses magical properties because it temporarily alleviates their sufferings of scabies and rashes (Braga 33–47). Then, Jameson identifies the "hermeneutic paradox Freud confronted when, searching for precursors of his dream analysis, he finally identified one obscure aboriginal tribe for whom all dreams had sexual meanings – except for overtly sexual dreams as such, which meant something else" (Jameson *Archeologies of the Future* 3). This may be a misquoting of Freud (I have not been able to locate it in *The Interpretation of Dreams* or elsewhere), but the point stands: dreams or texts about the climate crisis are as much about the desire, perhaps, for sublimation, qua creative response, another form of repression, perhaps. The Freud passage comes up early in Jameson's *Archeologies of the Future* in the context of a discussion of Ernst Bloch and utopianism; but in his earlier *Marxism and Form*, Jameson distinguished between what we might call a theological (or formal) narrative of Utopian wish, from how "the literary work attempts to use this Utopian material directly as content, in secular fashion, as in the various literary Utopias themselves" (Jameson *Marxism and Form* 145). Adapted to a climate reading, one might see the secular content being provided precisely by the kind of narratives that Ghosh demands. But in that case, what corresponds to the more formal questions of climate, to its theological dimension? Jameson locates that last tendency in the need for a "theory of figures," and that figural language was already attempted here in my discussion of Ruti's final essays.

While it is true that Ghosh is rather shooting fish in a barrel, attacking literary middlebrow fiction for not talking about climate change, it is interesting how he early on implicates his own practice, suggesting that it's something structural and historical, not a matter of *mauvaise foi*. Consider again Ghosh's demand, and the keywords "desires," "longings," and "yearning":

> Culture generates desires – for vehicles and appliances, for certain kinds of gardens and dwellings – that are among the principal drivers of the carbon economy. A speedy convertible excites us . . . because it evokes an image of a road arrowing through a pristine landscape . . . longings that are kindled in us have a chain of transmission that stretches back to Daniel Defoe and Jean-Jacques Rousseau . . . we are looking at an expression of a yearning that may have been midwifed by the novels of Jane Austen. The artifacts and commodities that are conjured up by these desires are, in a sense, at once expressions and concealments of the cultural matrix that brought them into being.
>
> (Ghosh 9–10)

Here I'd want to think about how desire works with respect to mimetic realism and literature: that is, desire of the Other. When we read Kerouac or Nabokov or Defoe or Rousseau or Austen, we desire what the novel desires. It's a wonderfully simple model that ignores different kinds of readers, different receptions (but then smuggling history back in as content or elided content as in the famous Said reading of Austen's *Mansfield Park*). And, too, to collapse those five authors (Kerouac, Nabokov, Defoe, Rousseau, Austen) into one literary category ignores national traditions (Ghosh is making strong claims throughout his book for Indian writers and how they deal, or do not deal, with climate) and flattens out the differences between, say, the layering of unreliable narrators in Nabokov from the innovation of Free Indirect Discourse in Austen and the proto-autotheory of Rousseau and the beatnik flow of signifiers in Kerouac. Then, what does that desire of the other mean (besides, again, extimacy)? As Alenka Zupančič argues in her recent *Let Them Rot: Antigone's Parallax*:

> What would it mean, therefore, from this perspective, to characterize Antigone's desire as "pure desire"? At first sight, it would seem that Antigone's desire is "pure" since, beyond or below all possible

metonymical objects, it aims at the Other's desire as such, at the fact that the Other is not only there but desires. The desire of the Other indicates that there is a point where the Other, so to speak, falls out of its own structure, does not cover it entirely; it shows to use the usual wording – that the Other itself is "lacking." But this lack in the Other is not the whole story; the lack itself has its obverse "positive" side: it functions as the entry point of contingency, and it has the power to transform what takes place at the level of this contingency into subjective necessity or destiny. "Pure desire" in this sense is not something abstract or purified of all particular objects; it affirms itself only with a particular or, rather, with a singular object. . . . Purity of desire is not contradicted by the existence of a singular object through which it affirms itself.

(Zupančič *Let Them Rot* 78)

What I take Zupančič to be arguing is that if the other desires (literature at large, or those specific novels or authors or characters) that means they also lack.

Overstory and creativity and suicide

In thinking of Ruti's concepts alongside the climate crisis, I draw also on one of the major psychoanalytic (but not, unlike Ruti's, in the Lacanian tradition) texts in this area, Renee Lertzman's *Environmental Melancholia: Psychoanalytic Dimensions of Engagement* (2015), which brings concepts of mourning and melancholia to bear on problems of environmental apathy and inaction. Lertzman asserts that there is an affective disconnect between our knowledge of climate disasterism and meaningful action, a disconnect we can compare to the barrier between learning/knowing climate change and an inability to write about it – the Ghosh thesis. However, as Nathan Gorelick points out, Lertzman, in neglecting to take drive and jouissance into account, may be underestimating how dire not only the actual conditions of climate disasterism are (which is to say, Ruti's contingent lack) but also the implications of those for the subject (grief, or anxiety), which are in turn related to or overdetermined by the constitutive lack in nature itself (Gorelick 222). Nonetheless, Lertzman may anticipate Ruti when she argues that effective climate activism needs to take creative forms, "how vital it is to invite creativity as a basis for engagement" (151). Richard Powers's novel demonstrates the stakes of such an investment in creativity. In the novel, a group of eco-activists, frustrated with logging practices, shift from nonviolent protest to sabotage, burning down mining camps, etc. (The novel's forbears include Edward Abbot's *The Monkey Wrench Gang*; a descendant is the film *How to Blow up a Pipeline*, based on Andreas Malm's political tract of the same name.)[1] After one of their number dies in an explosion, the group disbands, with one member, Nicholas Hoel, "making environmental art" ("The Overstory," Wikipedia) and another, Douglas Pavlicek, retreating into a ghost town cabin, writing a memoir, a "Manifesto of failure. Yellow legal-pad-paper pages pile scribbled in ballpoint pen pile up" (Powers 417). The memoir is discovered and read by a peripatetic hiker, who then trades that information to the police when she is arrested; Douglas, arrested in turn, betrays another member of the activist cell to federal authorities, on the sidelines of the Occupy protest at Zucotti Park in Manhattan. Douglas's memoir, his act of creativity, is both a metafictional stand-in for the novel itself *and* a demonstration of the limits, indeed the ethical disasters, of creativity. Am I arguing that Powers's novel shows, or "proves" the contradiction at the heart of Ruti's theory of sublimation? No. Or, rather, yes and no. First, yes, the narrative demonstrates not only that creativity can lead to a betrayal of the *polis*, of the activist cell, but that this literary device (the text within a novel, the writer as cutout for the novelist) is also necessarily a critique of Powers's own project.[2] But no, in that showing a contradiction in Ruti's theory is only to strengthen its critical possibilities.

For there is another political act (besides the dyad of sacrifice and sabotage, and then creativity proper), which is that carried out by the botanist scientist of the novel, Patricia Westerford, who is a stand-in (if the novel were to be read as a *roman-à-clef*) for Suzanne Simard, the Canadian author of *Finding the Mother Tree*, and the thesis that trees communicate (cue the Wittgenstein *bon mot* that if trees could speak to us, we wouldn't be able to understand them, and the more fundamental argument that this renders trees into speaking subjects, hence castrated by language) but also Peter Wohlleben, of *The Hidden Life of Trees*. At a certain point in the novel, Westerford, having tried to lend her scientific knowledge to efforts to stop deforestation and industrial logging practices, has a realization (like Douglas, who, in a bar one night realized that years of guerrilla-style tree planting was actually helping the forestry companies' reputations and pocketbooks), that is, seeing her work as locked into the university discourse, during a Ted talk–like event (such as Simard gives). Westerford commits suicide – the ethical act *par excellence* in the Lacanian reading of *Antigone*.

"Actually existing climate activism": helluva jouissance

I want to juxtapose these various forms of fictitious environmental activism in *The Overstory* – sabotage, writing, suicide – with the "Back to Fairy Creek" episode of *Helluva Story*, a radio/podcast series on the Canadian Broadcasting Corporation that revisits a protest against logging on Canada's west coast: the Fairy Creek occupation. Here we can think about both the populist frame of podcast qua genre (or medium or platform) and the trauma with which the activist confronts their lack. The signifier "helluva" is very much a sign of the popular, of the anti-elite vocabulary, not only in its meaning (which phrase, *helluva story*, denotes both a successful and/or efficacious narrative) but also a purely formal evaluation. The utterance speaks to the rhetorical skills of the teller, or perhaps the twists and turns the narrative takes. In this case, the "helluva story" is both that anti-logging activists realize their politics were, perhaps, for nought but also that they enjoy that realization. Moreover, "helluva" announces the mimicry of the oral: not three words, "hell-of-a," but one, run together. I should add that for my generation when growing up in a Christian Canadian family in the 1960s, we were not allowed to say "hell" – there is a slight *frisson* of obscenity in the phrase – we had to say – again, this was Canada, "h-e-double hockey sticks." So *helluva story* connotes a certain blue-collar, working class (masculine?) Canadian subject – best known, in popular culture of the 1980s as the Bob and Doug McKenzie "hosers" or today's TikTok explainers of Canadian slang.[3] But this subject coincides with the podcast qua genre, which as a communicative medium offers both the fantasy of the parasocial and the jouissance of the extimate: on the one hand, the parasocial is a form of transference, where we transfer our libidinal anxieties vis-à-vis the family romance or other proximate subjects (work- or schoolmates, strangers on the street, the neighbor) onto the podcaster or social media influencer; this then becomes a site for unbearable enjoyment (all enjoyment is unbearable: like food for an anorexic or bulimic, enjoyment for the Lacanian subject is either too much or not enough). The podcast frame, that is, overdetermines content: all podcasts are by their nature populist, are anti-elite, they are the terrain on which a struggle takes place now. In a useful commentary on an earlier draft of this chapter, Rosemary Overell makes the point that genre overdetermines ideology, as even highbrow cultural stalwarts like *Why Theory?* or *London Review of Books* move into chatty informality via the podcast.

What do we see in the "Back to Fairy Creek" podcast? A quick recap: the podcast is the story of Will O'Connell, a teacher on Vancouver Island, on Canada's west coast, who was involved in the largest act of civil disobedience in Canadian history, protesting the logging of an old growth forest in the Fairy Creek watershed. In the summer of 2020, conservationists catalogued the watershed after word spread that a logging company was building access roads in the area:

> Yellow-cedars are the longest-lived life forms in Canada, with the oldest one, located on the Sunshine Coast and cut down in 1993, recorded as being 1,835 years old. At 9.5 feet wide, the largest one we measured in the Fairy Creek headwaters could very well be approaching 2,000 years in age.
>
> (Ancient Forest Alliance)

Over 1,000 protestors would be arrested in the next year, marking the return of the so-called war of the woods of similar actions (the Clayquot protests) in the 1990s. The podcast concerns O'Connell's return to the site two years later, in spring 2023.

Here I want to point to two germane moments in the podcast: first, when O'Connell visits the site of the former blockade and there is "no remnant no physical memories" ("Return to Fairy Creek," 4 minutes: "I want there to be some memory of the blockade written into our legislation, not out here on the logging road"), and at another point ("Return to Fairy Creek," 18 mins) when O'Connell visits a site where activists were sitting in trees to stop logging, and now the entire cut has been logged, every tree is gone. It is difficult to determine what is more upsetting for O'Connell: the absence of trees, their ecological lack, or the slow rate of progress in policy and legislation. Nature, the trees, or the remnants of the blockade, that is, serve the function of the activists' sublime object, an object that they never had, which must be re-found, the lack of which object is mistaken for its loss. But to the detriment of the activists, the very frame of the podcast overdetermines the message – to Lacanize the Canadian media theorist Marshall McLuhan, the media qua Thing is the message qua lack.

The politics of the podcast affordances, that is, interpellate a subject who disavows environmental and other expert knowledge. Environmental knowledge is both that of the expert and of the activist: in Lacanese, both partake of the university discourse, or what the climate denialists, drawing on the playbook of Big Tobacco, call junk science and sound science (Oreskes and Conway). Here we should take account of an important Freudo-Lacanian argument with respect to jouissance and knowledge. Todd McGowan has provocatively argued that simple education will not work to eradicate racism, based as the latter is on jouissance, on the unconscious. In "'Wild' Psycho-Analysis," Freud writes:

> If knowledge about the unconscious were as important for the patient as people inexperienced in psychoanalysis imagine, listening to lectures or reading books would be enough to cure him. Such measures, however, have as much influence on the symptoms of nervous illness as a distribution of menu-cards in a time of famine has upon hunger. The analogy goes even further than its immediate application; for informing the patient of his unconscious regularly results in an intensification of the conflict in him and an exacerbation of his troubles.[4]

To which McGowan adds, "[i]f one tries to correct racist ideas by giving people better information, one inevitably fails because one does not touch the unconscious" (McGowan 193-194n). To adapt this formulation to our present topic, environmentalist education triggers insofar as it presents itself shorn of jouissance or ideology.

Jouissance, that is, is what lies beyond knowledge, which designates not so much truth (which is only ever uncovered, it seems, in the analytic session, but usually remains unconscious) as the "university discourse," the scientific register. Freud and McGowan's arguments are instructive, however, for they help us understand the climate denialists' alienation and their imperviousness to well-meaning education campaigns, whether of the top-down social media and state/Big Green varieties or the more peer-based friends and family or even doctor (nature as cure). Essentially, both forms of education or knowledge – through personal contacts or the medium of the digital – come wrapped up in the psychic resonances of enjoyment, anxiety, disavowal, foreclosure, and the like. This is to disagree in some ways with the claim that climate denialism is due only to the campaigns funded by Exxon, BP, or the

Canadian Association of Petroleum Producers. In a form of *coincidentia oppositorum*, the various post-structuralist and Lacanian theories of the decline of symbolic efficacy (there is no big Other, the signifier of lack in the Other, but also the postmodern end of *grand récits*) then meet, on the social media terrain, where the digital sublime means the affordance qua transferential parasocial. Think of this dynamic as a kind of information adultery: one's nagging wife or child or hectoring boyfriend or coworker are just a pain, whereas social media friends will always be there, the lure of the parasocial, the comforting murmur of the podcast voice in one's extimate AirBuds, which overdetermines any putative content (as in *Helluva Story*, where a climate activist content is negated). If, as McGowan and Ryan Engley recently remarked on the *Why Theory* podcast, knowledge is a barrier to enjoyment, then the opposite is also true: enjoyment, or jouissance, is a barrier to knowledge.[5]

What does this question of jouissance and the decline of the big Other mean for the climate subject? We know that climate denialism is not simply a left and right phenomenon as too easily read in the polarized discourse of U.S. politics (see, in this regard, Greg Garrard et al., *Climate Change Scepticism: A Transnational Ecocritical Analysis*). Perhaps we should think of the forms of negation at work in that very phrase "climate denialism." Who or what is denying or being denied, and what are the nuances?

Climate denialist as ethical subject

But it is also possible to see the climate denialist as an example of the ethical subject, who, committed to the singularity of their desire, does not give up on their jouissance, sees that there is no big Other, a radicality of subjectivity that should teach us something about our own political position. What do I mean by this? Consider the activism at work in *The Overstory*: the committed tree-sitting and other forms of obstructionism that we also saw in the Fairy Creek protests in British Columbia in 2020–2021 (indeed, Powers is on the record as admiring those actions). This is a scene from the tree-sitters' confrontation with a logging helicopter, combining *Apocalypse Now* and *The Monkey Wrench Gang*:

> The chopper is big, with a bay like a bungalow. Big enough to hoist a tree older than America straight into the sky and haul it upright

across the landscape. Its blades froth the air around the dangling girl. Two humans sit inside the fiberglass pod, cloaked in visors and chin-cupping helmets, chatting on tiny boom mics with some distant mission command.

Adam stares at the trick of blockbuster back projection. He has never been so close to a thing so huge and malevolent. He sees its million parts – shafts, cams, blades, plates, things for which he doesn't even have a name beyond the power of any human to assemble, let alone design. Yet there must be thousands of such craft, employed by industries on every continent. Tens of thousands more, armed and armored, in the globe's many arsenals. World's most common raptor.

Branches snap off and the air fills with chaff. Burnt fossil steams from the beast, stinking like a burning oil rig. The stench gags Adam. The roar pierces his eardrums, killing all thought. The woman flaps on her branch like a pennant, then drops her weapon and holds on. Her filming partner loses his grip in the artificial gale, and the camera, too, drops two hundred feet and plinks apart. A metallic voice, massively amplified, comes out of the helicopter. *Exit the tree, immediately.*

(Powers 354)

The scalar disjunction between eco-activists in the tree and malevolent machine brings to mind the *non-rapport* of asymmetrical warfare, or Adorno's negative observation that street protests are helpless before ICBMs. (And yet are both tree and chopper not sublime objects?) Or a temporal cut, cut meant both in the Lacanian sense (is the logging chopper's command not the voice of the analyst, the desire of the Other?) and in the extractive violence of tree farming. Consider another scene of a less apocalyptic, more midcentury *soignée*. Here I am thinking of the redwood forest scene in Hitchcock's *Vertigo*.

But I compared the tree-sitting scene in Richard Powers's novel to its antecedent in Edward Abbey: that is not right. For the sabotage that characterizes *The Monkey Wrench Gang* (or its real-life successors the activist group Earth First!) actually occurs later in *The Overstory*, when, peaceful protest having failed, a small cadre turns to property-destroying violence. Here is Andreas Malm describing a similar action when, in 2016, the activist group *Ende Gelände* ("Here and no further") confronted the Schwarze Pumpe coal facility in Germany:

[W]e encountered a fence. Walking, half-running in the front, my affinity group tore it down, broke it apart, stamped on it and continued with the rest of the march up to the perimeters of the plant. . . . The few private guards caught off-hand and completely outnumbered, we rushed into the compound. During my years in the climate movement, I have never felt a greater rush of exhilaration: for one throbbing, mind-expanding moment, we had a slice of the infrastructure wrecking this planet in our hands.

(Malm 159)

In the following chapter, Žižek's version of Kübler-Ross's stages of grieving and their supplement with Mann's stages of climate denialism will help us to understand some of how to think about the politics of climate activism in terms of Mari Ruti's theory of sublimation. But here, recall Ruti's argument that the fall from fantasy, the disenchantment of our precious signifiers, our libidinal treasures, can actually spur us to more profound relations to the singularity of our desire.

[I]t is only the fall of our most treasured fantasies – particularly of the idea that there is some "sovereign good" that is capable of shielding us from the terror of living – that allows us to transition to a more imaginative and creatively engaged psychic economy.

(Ruti, "The Fall of Fantasies: A Lacanian Reading of Lack" 486)

This fall from fantasy triggered by the plunder of our libidinal treasures helps us to understand the *coincidentia oppositorum* of anthropogenic climate change with climate denialism. As Steven Swarbrick argues, the climate crisis constitutes a traversing of the fantasy of nature as the big Other, the guarantee of our subjectivity "becom[ing] meaningful only when living systems do not work – that is, when life is no longer *for us* but involves the suspension or dissolution of the actual as such" (Swarbrick 224). This traversing surely is responsible for the jouissance at work in Malm's narrative but also in that confrontation with the helicopter in *The Overstory*. In Chapter 1, I mentioned contradictions in Mari Ruti's work, including what I take to be a mistaken theory of how lack accumulates, its fungibility – this is not to see lack in terms of its irreducible negativity (we also must be wary of the ontologization of lack, of the non-rapport – as Krzych has pointed out, Ruti's concept of the object or thing differs dramatically from Žižek's theory, which "emphasizes the horrifying features of the Thing as a *sublime* object" albeit with an emancipatory potential). And, as Ruti

remarks in her essay on the fall from fantasy, that fantasy differs signif-
icantly from the fantasmic object as *sinthome* worked out in *The Singularity of Being*. Indeed, this call for the singularity of desire is why
in some ways we have to recognize that the climate denialist, *like the
activist*, is an ethical subject: confronted with the lack in the big Other
of the university discourse, they have remained true to, they have re-
fused to cede, their desire. The climate denialist sees that there is no
other of the Other (as per Kryzch – there is no guarantee of the ring or
the tree) but is also a pervert who, like Ruti, knows very well (hence
the fetish). Perhaps a formulation devised many years ago by Claude
Lévi-Strauss (and discussed with respect to *Antigone* in Chapter 2) will
help us. In the A:B::C:D structure, it is not so much we cannot see
the tree for the forest, but we cannot see the ring for the tree, and so
ring:tree::tree:forest.

What of the negative subject? Here we can expand the queer subject, Ruti's purview, with the permission of Jasbir Puar's project, which
reads terrorist suicide bombers as queers. But I want to do so in a way
that, in spite of its subjectless intentions, makes space for such precarious subjects as the junkie, the unhoused, missing and murdered Indigenous women and girls and 2spirit people, the precariat, but also the
carbon subject, the Trumpists and populists, the COVID-19 skeptics
and anti-vaxxers. I want to pose those subjects in terms of a variety of
taxonomies of psychoanalytic subjectivities: the varieties of disavowal
and denialism (and perversion and splitting) as developed by Hegel,
Freud, Žižek, and Zupančič; Freud's "kettle logic" or the argument
that the unconscious brooks no contradiction; Hegel's "beautiful soul"
and Melville's Bartlebyan "I would prefer not to," and the five stages
of grief promulgated by Elizabeth Kübler-Ross and also taken up by
the climate scientist Michael Mann. My aim is to suggest that a way of
reading Ruti is in terms of an emancipatory theory of grief.

Freud's four "Ver-"s with Hern and Johal

In *Mourning Sickness*, Rebecca Comay tells us:

> Hegel repeatedly anticipates Freud's terminology as he investigates
> enlightenment's inquisitorial agenda: disavowal (*Verleugnung*),
> perversion (*Verkehrung*), splitting (*Trennung, Entzweiung*), isola-
> tion (*Isolierung*), the stubborn forgetting (*Vergessen*) of the lost

object – the catalogue details the defensive apparatus of a subject bent on sustaining itself on what it gives up.

(64–65)

And in *Less than Nothing*, so Žižek:

> [I]n Freud there are four main forms, four versions, of "*Ver-*": *Verwerfung* (foreclosure/rejection), *Verdrängung* (repression), *Verneinung* (denial), *Verleugnung* (disavowal). In *Verwerfung*, the content is thrown out of the symbolic, de-symbolized, so that it can only return in the Real (in the guise of hallucinations). In *Verdrängung*, the content remains within the symbolic but is inaccessible to consciousness, relegated to the Other Scene, returning in the guise of symptoms. In *Verneinung*, the content is admitted into consciousness, but marked by a denial. In *Verleugnung*, it is admitted a positive form, but under the condition of *Isolierung* – its symbolic impact is suspended, it is not really integrated into the subject's symbolic universe. Take the signifier "mother": if it is foreclosed or rejected, it simply has no place in the subject's symbolic universe; if it is repressed, it forms the hidden reference of symptoms; if it is denied, we get the by now familiar form "Whoever that woman in my dream is, she is not my mother!"; if it is disavowed, the subject talks calmly about his mother, conceding everything ("Yes, of course this woman is my mother!"), but remains unaffected by the impact of this admission. It is easy to see how the violence of exclusion gradually diminishes here: from radical ejection, through repression (where the repressed returns within the symbolic) and denial (where the denied content is admitted into consciousness) to disavowal, where the subject can openly, without denial, talk about it.

(859, chapter 13)

With *Verwerfung* (foreclosure/rejection), the hallucination returns in the form of paranoid conspiracies, such as the certainty that forest fires in Greece were caused by arsonists in the pay of the World Economic Forum as part of the "Great Reset" (as a friend said to me while on a hike a few summers ago) and not, in fact, global warming, the drying of underbrush, increased lightning and wind speeds – the conditions of "21st-century fire" that John Vaillant discusses in *Fire Weather*. That is, climate change or global warming has no place in the subject's symbolic universe. With *Verdrängung* (repression), the content is in the symbolic but inaccessible and only surfaces as symptom: climate grief, climate anxiety (or even climate love): here we first have to recognize that climate affects are often a screen. We may seem to be troubled by the smell of forest fire smoke or the appearance of microplastics, but that is just a screen over grieving for one's dead relative – or the other way around.

In terms of *Verneinung* (denial), what appears to be the template for climate denialism itself bifurcates: this is actually the second stage of denialism, the labeling of climate science as "bad science" or "junk science," but this is because "the denied content is admitted into consciousness" – first, that is, historically (and this is the playbook that the climate denialists inherited from Big Tobacco), there was a form of disavowal, or *Verleugnung*, it is admitted as positive form, but under the condition of *Isolierung* – its symbolic impact is suspended, it is not really integrated into the subject's symbolic universe. There will be the call for more research from tobacco or climate institutes, waffling on how there is not consensus and the like. Importantly, while research qua research is positive (positive *form* is a version of negation), it is also isolated or segregated. Indeed, the positive operates qua fetishization, and denialist's organizations often emulate legit ones. Thus, David Lipsky notes, tobacco lobbyists created the Advancement of Sound Science Coalition, a near-copy of the American Association for the Advancement of Science (Lipsky 343); infamous climate denier Fred Singer coined "the NIPCC Report – the Non-Governmental International Panel on Climate Change. . . . it sounded just like [the IPCC,] the United Nations' Intergovernmental Panel [on Climate Change]" (Lipsky 276–277). Fetishistic disavowal works both at the level of the signifier and in terms of substance (sound science, climate change). Here my claim that we are all climate denialists is borne out, for of course those of us who dutifully sort our plastics and paper into recycling bins, or purchase carbon offsets when we fly, *we know very well that recycling does sweet fuck all*, but we feel better when we do it.

We can think in a schematic way of how this spectral analysis of the "*Ver-*"s and the denials, the disavowals, and the repression work in terms of climate change. It is not very hard for a university professor like myself to disavow my role in climate change by fetishizing my so-called knowledge (yes, I know, the earth is warming up – as Sheila Heti says in this book's epigraph, but you know, I take public transit, or recycle, or I bought carbon offsets for my six airplane trips in the last year). But, to move quickly to the other end of the spectrum, for my relatives who work in the oil and gas industry, their relation to negative subjectivity is much more fraught – and direct. Their attitude is more likely to be approaching foreclosure, a kind of psychosis that explodes in anger at my liberalism, or, as in the social media interactions I had when at a Petrocultures conference in Edmonton, in 2012, I referred to oil workers as "rig pigs" (I never knew resource workers were such snowflakes).

But to move out of my own narcissistic autotheory, consider Matt Hern and Am Johal's book *Global Warming and the Sweetness of Life: A Tar Sands Tale*, in which, with cartoonist Joe Sacco, the activists take a series of road trips from Canada's west coast to Fort McMurray, interviewing oil patch workers, Indigenous residents, and other native informants. Here we can see forms, for instance, of denial: so Vanessa, a brew pub worker tells the authors: "But everyone there [Fort Mac] is so hyperaware of the environment and so hyperaware of being judged, you form a bubble around yourself. I feel fine about it. If I wasn't there, someone else would have taken that job" (Hern and Johal 164), and Gita, also worker at a brew pub: "The scale of pollution and environmental issues are so huge, how can that be my responsibility, or my fault?" (Hern and Johal 105).

And repression makes its effect known by the severity of symptoms, in particular of drug use and violence against women:

> For years . . . crime statistics across the board reflected boomtown pathologies of drugs and violence, assault and alcohol. The stories get amplified and exaggerated for more genteel ears down south, but the reality was (and in many ways still is) that the tar sands are a tough place to earn a living, and especially so for women. The violence in the Wood Buffalo region peaked in 2008, though, and through 2015 declined on almost all counts, reflecting comprehensive policing and policy efforts. Starting in 2015, and hardly surprisingly mirroring the downturn in oil prices, violence statistics started to turn upward again.
>
> (Hern and Johal 103)[6]

This is all to problematize the question of denialism as a response to the lack or trauma that is climate change.

Salina et al.: semiotic rectangle 2

Hegel is well known for taking the tools of the Enlightenment critique of religion as a way to pull the rug out from under Enlightenment philosophy:

> For Enlightenment does not employ principles peculiar to itself in its attack on faith, but principles which are implicit in faith itself. Enlightenment merely presents faith with its *own* thoughts which faith unconsciously lets fall apart, but which Enlightenment brings together; it merely reminds faith when one of its own modes is present to it, of the others which it also has, but which it always forgets when the other one is present. Enlightenment shows itself to faith to be pure insight by the fact that, in a *specific* moment, it sees the whole, brings forward the other moment which is opposed to it, and, converting one into the other, brings to notice the negative essence of both thoughts, the Notion. To faith, it seems to be a perversion and a lie because it points out the *otherness* of its moments; in doing so, it seems directly to make something else out of them than they are in their separateness; but this "other" is equally essential and, in truth, is present in the believing consciousness itself, only this does not think about it, but puts it away somewhere. Consequently, it is neither alien to faith, nor can faith disavow it.
>
> (Hegel, *Phenomenology*, §564, pg. 344)

If we can see a number of prescient analytic moments in Hegel, thanks to Comay's identification of the proto-psychoanalytic mechanisms at work in the *Phenomenology*, how do they help us understand the grieving subject of climate change? Here some of the key mechanisms include the unconscious, of course, as well as perversion, the lost object, and disavowal.

I want to explore these logics via a subject who seems very much not to be a grieving subject, the artists who make work out of the garbage of late capitalism: Tita Salina's *1001st Island* (a work in which an island is constructed from floating garbage in the Jakarta harbor). Salina's work very much demonstrates the ambiguous way in which we confront the lost object of nature in its evil other, garbage or pollution. Our grief in the face of climate change is a mourning for an object we never had, a mourning that, transmogrified into melancholy, tosses back and forth like a sleepless dreamer, first fetishizing the forest or ocean or charismatic megafauna, then vilifying plastic garbage or neon-hued tailing pond or benighted petro-prole. Salina's work is a video that begins with the scooping up of floating waste in the Jakarta harbor. Garbage in the Global South is nothing new and has everything to do with how that hemisphere has, since the 1980s, been the destination for most of the developed world's toxic, odoriferous, or simply "forever" waste, from electronics crushed into nothing in Ghana, to murderous plastic fires on the byroads of Cambodia (Clapp). The *1001st Island* constitutes itself as a work of social relation from the start – fisherfolk and other harbor denizens take part, this is all very charming and reassuring. Third world subjects are active here, they have agency.

Yes, we know this is artwork (I first watched Salina's work at the Hamburger Bahnhof in Berlin, a repurposed train station that is one of the major German sites for contemporary art), and that the Jakartans are doing this under the direction of the artist, but we can still have our fantasies, can we not? The harbor is still getting cleaned up, regardless. But at a certain point this stops, and we next see the plastic waste being assembled into some kind of a flat contraption, which is then carried on the shoulders of many of the same workers. (Is the artist in there as well? We would feel better if she is, or perhaps the opposite is also true, and we would feel better if she was not, if she maintained her autonomy, didn't try to disavow social hierarchies.)

In this way, Salina's work stands as a riposte to the previous (1990s and 2000s) mania for "relational aesthetics," such as Rirkrit Tiravanija and other, as one vulgate had it, "artists-who-serve-soup-at-the-opening" (Bourriaud 7). Unlike that moment, with Salina we have not food being served but garbage scooped out of the harbor (like a soup, I suppose), but also, as with Tiravanija, "lots of people" and in fidelity to the relational ethos of sociality, as well as the discovery in the 1990s of the thrift store and flea market as a source for art, further downgraded, to the (floating) garbage heap (Bourriaud 47). Now this flat mass of waste is brought into the water and towed out of the harbor. At a certain point the artist stands upon it herself, dressed in black, stands and then lies down, and here a couple of decisions or options present themselves. On the one hand we have the artist associated with trash, with the trash that, *à la* great Pacific garbage patch, has come to be the master signifier of global pollution. This *1001st Island* then is somewhere between that gigantosaur and its scalar other, the microplastics and nanoplastics that are to be found in sea creatures' stomachs, in our testicles, in our water and air.

Such an artwork, trash made into an island, suggests something both more and less than garbage. It is more than garbage in the sense that it is now part of a social artwork, rescued from simply being trash in the harbor, and now possessing some aesthetic value. But it is also less than garbage insofar as it is an island. Are not such artificial islands, such as appear off the coast of the United Arab Emirates, or are the domain of "seasteaders" or are analogous to the reclaimed lands of Singapore and other coastal communities (as shown in the Singapore film *A Land Imagined*), in their fey attempts to emulate nature, and indeed in their own antinomic relation with the disappearing coastline ("Toronto may well be a coastal city one day" [Heti 190]), not so much the heroic villains of trash but merely some wannabe symptom of the coming soggy apocalypse?[7]

To help us understand these subjectivities at work, especially with respect to climate activists, resource workers, but also the artist, I conclude this chapter with the algorithm popularized by Fredric Jameson, the semiotic rectangle, Figure 3.1. This offers us the antinomy of the artist or the subject who does not seem to be grieving (who at least does not perform that affect in a legible way) with the trash island, and then spins off various oppositions and opposites.

This diagram is discussed more fully in the next chapter, where I situate Salina's art in terms of other artists working with trash (including Adán Vallecillo, whose *Saturación* I mentioned in the introduction). But it is worth, perhaps, returning to some of the arguments in this chapter, both with respect to the question of climate fiction as well as a

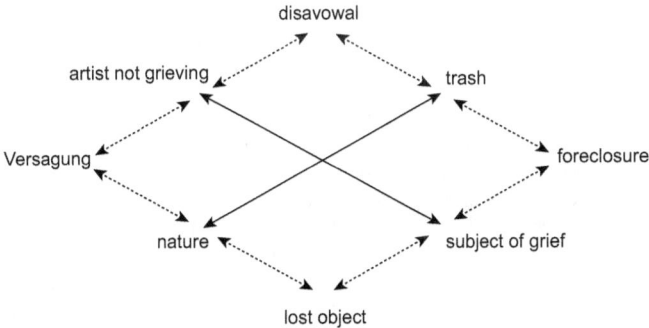

Figure 3.1 Semiotic rectangle of climate grief

psychoanalysis of denialism. These conundrums – the Ghosh paradox whereby the novel causes climate change but cannot depict it – and literature or art as a form of sublimation in the sense that Ruti means in her late writings – help us to think about the fungibility of lack and the politics of grief, which is to say, an emancipatory theory of lack.

Notes

1 I am grateful to Kelly Gray, PhD student at Boston College, for directing me to Powers's novel and its Lacanian overtones.
2 There is also a way in which creativity can traverse the fantasy of politics, of the dissolution of the social. In the activist zine *Creeker Companion*, published by members and fellow travelers of the Fairy Creek protests, advice is given on not talking to the police; the zine also prints a historical account, from the "Twin Cities General Defence Committee," of the role of "bad jacketing" or spreading disinformation on activists to undermine their credibility. See, for the origins of the term "bad jacketing," Churchill and Vander Wall, 49–51.
3 See "Canadian Slang You Need to Know."
4 Freud, " 'Wild' Psycho-Analysis," 225. Ctd. McGowan, *The Racist Fantasy*, 193–194n. The menu trope appears to have come from the 18th-century satirist Georg Lichtenberg, according to Walter Lowrie, perhaps via Kierkegaard on Hegel: "he [Kierkegaard] quoted with appreciation Lichtenberg: 'It is about like reading out of a cookbook to a man who is hungry'" (Lowrie 115).
5 Ryan Engley and Todd McGowan, *Why Theory*. Podcast. Web. 16 September 2023.
6 The authors also note:

> Fort Chip is 220 kilometers north (downstream) from Fort Mac. It is a town of about a thousand, predominantly Cree, Chipewyan (Dene), and Métis people that is wracked by awful disease and cancer rates. In 2014, Health Canada finally recognized what residents had known for decades. In the words of one researcher: "Something unique is happening in Fort Chipewyan, especially around cancer." The three-year million-dollar study interviewed 94 people and reported 23 cases of cancer, linked directly to high levels of contaminants from bitumen extraction found in the fish and animals residents rely on for food. The report also noted a wide range of other diseases, confirming the findings of multiple previous studies, and linked all directly to tar sands production.
>
> (Hern and Johal 34)

7 The "Tropical Imaginaries and Climate Change" issue of *eTropic: electronic journal of studies in the tropics* (2021) contains important work not only on decolonizing climate change (Chao and Enari 2021) and the relation between ecocriticism and postcolonialism (Harnett 2021) but also, germane to Salina's work, the speculative fictions " 'The Post-Qantal Garden' Annotated" (Boswell 2021) and "Goodbye on the Seas: Rising Waters, Submerging Lives" (Yin 2021).

References

Ancient Forest Alliance. "Massive Old-Growth Yellow Cedars, Including Canada's Ninth-Widest, Under Threat in One of Vancouver Island's Last Intact Valleys." *Media Release*, 13 August 2020. https://ancient-forestalliance.org/media-release-fairy-creek/. Accessed 22 May 2025.

Boswell, Jake. "The 'Post-Quantal Garden' Annotated." *eTropic: Electronic Journal of Studies in the Tropics* (2021). http://dx.doi.org/10.25120/etropic.20.2.2021.3817.

Bould, Mark. *The Anthropocene Unconscious: Climate Catastrophe Culture*. London: Verso, 2021.

Bourriaud, Nicholas. *Postproduction*. New York: Lukas & Sternberg, 2002.

Braga, Rogelio. *Is There a Rush Hour in a Third World Country?* London: the87press, 2022.

Burnham, Clint. "Lacan's Trash Talk: Three Objects for the Internet." In Burnham and Kingsbury, Eds., 75–93.

Burnham, Clint and Paul Kingsbury, Eds. *Lacan and the Environment*. London: Palgrave, 2021.

"Canadian Slang You Need to Know." *TikTok Video*, 18 April 2023. https://www.tiktok.com/@heppellspotato/video/7223506336624020741?lang=en. Accessed 22 May 2025.

Chao, Sophie and Dion Enari. "Decolonising Climate Change: A Call for Beyond-Human Imaginaries and Knowledge Generation." *eTropic: Electronic Journal of Studies in the Tropics* (2021). http://dx.doi.org/10.25120/etropic.20.2.2021.3796.

Churchill, Ward and Jim Vander Wall. *Agents of Repression: The FBI's Secret Wars Against the Black Panther Party and the American Indian Movement*. Cambridge, MA: South End P, 2002.

Clapp, Alexander. "Everything You've Ever Thrown Away Is Likely Still Out There." *New York Times*, 16 February 2025.

Comay, Rebecca. *Mourning Sickness: Hegel and the French Revolution*. Stanford: Stanford UP, 2010.

Creeker Companion. *Zine*, 24 May 2023. https://creekerzine.wordpress.com/. Accessed 22 May 2025.

Freud, Sigmund. "Wild Psycho-Analysis." *The Standard Edition of the Complete Psychological Works of Sigmund Freud*. 1910. Trans. Joan Riviere. Volume 11. Ed. James Strachey. London: Hogarth P, 1957. 219–227.

Garrard, Greg, Axel Goodbody, George B. Handley and Stephanie Posthumus. *Climate Change Scepticism: A Transnational Ecocritical Analysis*. London: Bloomsbury, 2019.

Ghosh, Amitav. *The Great Derangement: Climate Change and the Unthinkable*. Chicago: U of Chicago P, 2017.

Gorelick, Nathan. "Psychoanalysis at the End of the World." In Burnham and Kingsbury, Eds., 221–238.

Hern, Matt and Am Johal. *Global Warming and the Sweetness of Life: A Tar Sands Tale*. Cambridge: MIT P, 2018.

Heti, Sheila. *Alphabetical Diaries*. Toronto: Penguin Random House, 2024.

Jameson, Fredric. *Archeologies of the Future: The Desire Called Utopia and Other Science Fictions*. London: Verso, 2005.

Jameson, Fredric. *The Geopolitical Aesthetic: Cinema and Space in the World System*. Indianapolis: Indiana UP, 1995.

Jameson, Fredric. *Marxism and Form: 20th Century Dialectical Theories of Literature*. Princeton: Princeton UP, 1971.

Krzych, Scott. "Circumstantial Sublimation and Steven Soderbergh's *Ordinary Objects*." *Psychoanalysis, Culture & Society* 23.2 (2017): 123–140.

Lacan, Jacques. *The Seminar of Jacques Lacan: Seminar VII, 1959–60, The Ethics of Psychoanalysis*. 1986. Trans. Dennis Porter. New York: Norton, 1997.

Lacan, Jacques. *The Seminar of Jacques Lacan: Seminar XI, The Four Fundamental Concepts of Psychoanalysis*. Trans. Alan Sheridan. New York: Norton, 1998.

Lertzman, Renee. *Environmental Melancholia: Psychoanalytic Dimensions of Engagement*. New York: Routledge, 2015.

Lipsky, David. *The Parrot and the Igloo: Climate and the Science of Denial*. Toronto: Penguin Random House, 2023.

Lowrie, Walter. *A Short Life of Kierkegaard*. 1942. Princeton: Princeton UP, 2013.

Malm, Andreas. *How to Blow up a Pipeline: Learning to Fight in a World on Fire*. London: Verso, 2021.

McGowan, Todd. *The Racist Fantasy: Unconscious Roots of Hatred*. New York: Columbia, 2022.

Moylan, Tom. *Scraps of the Untainted Sky: Science Fiction, Utopia, Dystopia*. New York: Routledge, 2000.

Oreskes, Naomi and Erik M. Conway. *Merchants of Doubt: How a Handful of Scientists Obscured the Truth on Issues from Tobacco Smoke to Global Warming*. New York: Bloomsbury, 2010.

"The Overstory." *Wikipedia.* https://en.wikipedia.org/wiki/The_Overstory. Accessed 22 May 2025.

Powers, Richard. *The Overstory.* New York: Norton, 2018.

"Return to Fairy Creek." *CBC.* Podcast, 21 June 2023. https://www.cbc.ca/player/play/audio/1.6876515. Accessed 22 May 2025.

Ruti, Mari. "The Fall of Fantasies: A Lacanian Reading of Lack." *Journal of the American Psychoanalytic Association* 56.2 (2008): 483–508.

Ruti, Mari. *The Singularity of Being: Lacan and the Mortal Within.* New York: Fordham UP, 2012.

Swarbrick, Steven. *The Environmental Unconscious: Ecological Poetics from Spenser to Milton.* Minneapolis: U of Minnesota P, 2023.

Szeman, Imre. "Introduction to Focus: Petrofictions." *American Book Review* 33.3 (2012): 3.

Vaillant, John. *Fire Weather: A True Story from a Hotter World.* Toronto: Penguin Random House, 2023.

VanderMeer, Jeff. "Climate Fiction Won't Save Us." *Esquire,* 19 April 2023. https://www.esquire.com/entertainment/books/a43541988/climate-fiction-wont-save-us/. Accessed 21 May 2025.

Yin, Christina. "Goodbye on the Seas: Rising Waters, Submerging Lives." *eTropic: Electronic Journal of Studies in the Tropics* (2021). http://dx.doi.org/10.25120/etropic.20.2.2021.3818.

Zupančič, Alenka. *Let Them Rot: Antigone's Parallax.* New York: Fordham UP, 2023.

Chapter 4

"Spectral analyses"

More on semiotic rectangle of climate grief

To recap, in Figure 4.1 we have two concepts that form an antinomy – the artist (who is not grieving) and trash. The negation of that artist would be the grieving subject, who mourns the loss of nature. Then, the opposite of trash perhaps is nature itself – in the banal idea that one should "pack out" any trash one creates when on a hike or camping trip, or the more egregious pollution of the wilds or the ocean. The four terms at the outermost points then combine those four concepts in a combinatory logic.

Begin at the top of the algorithm, with the subject who does not grieve, in this case the artist Tita Salina, who insouciantly stands on top of an island of trash. Now, this is an antinomy because evidently trash is not art in a conceptual or ideological sense. Art makes a claim to be something autonomous from the world, something of elevated aesthetics. Even art that, like this work I am discussing here, is physically or materially made from trash, nonetheless is a form of sublimation. And I mean this of course in the strictest Lacanian sense, returning to that figure, in *The Ethics of Psychoanalysis*, of the poet Jacques Prévert making a decorative chain around his apartment from matchboxes. Why, then, is what joins or resolves this antinomy a matter of disavowal? Am I saying that sublimation, raising the object to the dignity of the Thing, is also a kind of fetishism? I come back to that question but for now point to the importance of the artist qua subject, as performed or cosplayed by Salina in the video, in her black garb. As I said earlier, a couple of decisions or options present themselves when we see this figure. She is not dressed in any kind of ethnic regalia, for one, so this is not an "artist from the Global South" qua ethical subject. She is not making claims, at least not in a legible fashion, for her national or racial or ethnic authenticity; on the contrary, she is situating herself

DOI: 10.4324/9781003518914-5

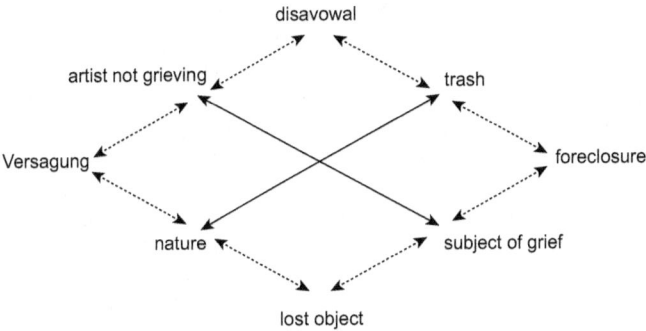

Figure 4.1 Semiotic rectangle of climate grief

in the global art world, discourse, or market. She is also saying she is not trash – which is not to say that the ethnic subject is trash, but rather that, even if she works with trash, she is nonetheless an artist. But in Salina's case, there nonetheless is textbook sublimation going on, as with other artists whose material is trash or garbage – Canadian photographer Kelly Wood with the *Continuous Garbage Project* (1998–2003), Dieter Roth with decades' worth of detritus, Yuji Agematsu, who "[e]ach day . . . finds objects while walking the city and drops them into the cellophane wrapper from a cigarette pack,"[1] Dawn DeDeaux with broken glass from New Orleans' Hurricane Katrina repurposed to look like the satellite image of a hurricane's vortex, Ghanaian artist Serge Attukwei Clottey with repurposed plastic Kufour canisters, and Adán Vallecillo's *Saturación*, the minimalist sculpture made of used oil filters we encountered in this book's introduction.

These various artists' sublimation as a response to the climate crisis does two further things: first, it brings together the two examples of Lacan – the canonized art practice of Cézanne, and the accidental or contingent garbage of Prévert. But it also raises the question of whether such creativity or practices in general constitute a kind of fetishist disavowal of that crisis or lack. That critique has been argued recently with respect to the British psychoanalyst Marion Milner by Clair Wills, who is suspicious of how "comforting [it is] to think that what is real in living lies in the basic creativity which we can all learn to call up" and of how the "joy in making things out of what we find is sober, cheap, easy to access and available to everyone," wondering if

Milner's " '[a]bsent-mindfulness' might be the perfect quietist therapy for today, a lucrative branch of the wellness industry" (Wills 6). Here two points must be made in response. First, Ruti's notion of creativity is based more on rather radical concepts, derived as much from Nietzsche as from Lacan, that creativity is a profound response to one's lack – and not a disavowal. Second, the role of artists toward garbage and trash can in the end be thought of neither as a form of social art (especially when carried out in the Global South, where much of the world's garbage ends up) nor as a literal example of an intervention into the despoliation of nature. I mean, yes, it is those things, but it is also taking an article that was not art and making it into art. This is sublimation no less for an apple (or a ring) than for a matchbox (or plastic detritus in the ocean). Sublimation is a formal act, not tied to the content or substance of the object.

This conundrum recalls the tragedy on offer in *The Cleaners*, a documentary about social media moderators working in the Philippines (Riesewieck and Block). Here one such moderator tells us she grew up next to a garbage dump in Manila Bay, and got an education to avoid, as her mother told her "if I don't study well, I'll end up as a scavenger" (Burnham, "Lacan's Trash Talk" 82). Of course, now she is tasked with cleaning up the internet. There is also a global network of such scavengers, recyclers, and the like, often self-organized like the Binners Project in Vancouver.[2] Now, evidently, artists who work with garbage, social media clean-up crews, and the urban poor who scavenge or dumpster dive occupy different social positions, so what does it mean to think of them at the same time? Am I saying the person who picks up empty cans outside my house is the same as an artist? Yes – but not because they are an artist, but rather that their practice is, too, a matter of sublimation.

From grief to anger: Kübler-Ross

To move on, on the right-hand side of the diagram we have trash aligned with the grieving subject. Here the subject's emotions or affects would range across a variety of positions, anger or hypochondria perhaps, but I want to suggest those associated with psychosis and foreclosure such as anger and paranoia. Not for nothing is anger the second of Kübler-Ross's five stages of grief. Freud's *Verwerfung* signifies foreclosure/rejection: grief takes these forms as a way of (r)ejecting trash, pollution, garbage from consciousness. Trash is a kind of social shit, and we are all stuck in toilet training. The psychic response to trash is very much like, is a kind of repetition of, the social response at large: we want to get rid of it. Here we do not merely repeat the *coincidentia oppositorum* of climate denialism and climate grief argued earlier. Foreclosure is a stronger rejection than that of denial, which in its canonical form takes the expression "whoever that woman is in my dreams, she is not my mother!" Denial places the subject in the dyad of the *sujet d'énoncé*, the speaking subject, as opposed to the *sujet d'énonciation*, the subject of the unconscious, revealing the contradiction at work. Recycling is the denialist position *par excellence* insofar as the very busywork of sorting recycling gives one the illusion that this is not trash.

Be that as it may, foreclosure's violence, its anger, is very much what is available to the grieving subject because of the narcissism of grief. As Žižek puts it in his application of Kübler-Ross, anger occurs "when we can no longer deny the fact: 'How can this happen to me'?" (*Living in the End Times* xi). Why me? Anger is a symptom of foreclosure and a response to grief. In Žižek's utterance, there are three germane moments: denial has been tried out, the subject is affronted that they are affected by a climate disaster, and thence a turn to violent anger. Now, many critics, including Kübler-Ross, have argued that the stages of denial, anger, bargaining, depression, and acceptance are not a simple linear path, and that the subject of grief will often cycle through, or skip, or repeat, the stages. But the difference of anger from denial, and its following of denial, is important. Foreclosure, like disavowal, is not denial. And violent anger seems to be necessary psychically to remove the specter of whatever it is that must be foreclosed. Trash itself is a "bad object" as well as signifiers or discourses of same, ranging from so-called toxic masculinity to government regulations regarding plastic bags, paper straws, or electric vehicles.

Anger and foreclosure thus are not simply responses to grief but substitutions, compromise formations. Climate grief is a matter of not having realized one's desire: grief is in effect, in this reading, a screen for guilt, as is made evident in Žižek's formulation that anger occurs as a response to grief but also recalls Lacan's "the only thing of which one can be guilty is of having given ground relative to one's desire" (*Seminar VII* 319). Here the contradiction of the singularity of one's desire versus the extimacy (desire is the desire of the Other) is coming to its fore. So, one's reaction to climate change begins as that guilt founded on the impossible choice of not ceding one's desire when it isn't exactly clear what one's desire is. Do I want to enjoy my SUV, for all its petro-pollution or self-righteously sort my recycling? Do I support jobs for Indigenous and non-Indigenous workers, or is my desire to stop pipelines, and stop their propensity for spills? But that guilt then is obscured or swapped in by grief as a screen, a compensation or compromise. In part this is because of the structural nature of desire: we desire to desire, and its object (SUV, recycling, solidarity) is unimportant to our affect. Grief is more enjoyable, and compensates for the lack of the solidarity, or consumer goods, we were unable to obtain. But grief sets off its own unpredictable (if for the subject) chain of denial, anger, and so on. And just to be clear, this analysis is not suggesting some Leninist or Lacanian "subject supposed to know" more than the suffering subject of climate grief.

Graphs of acceleration: "dick pics" of lack

Circling to the bottom of the graph, this same subject of grief will also have a different range of emotions with respect to nature itself, figuring nature as a lost object. Here we have first to state the orthodox Freudian position that the lost object is always an object of retroactive loss, is created by the loss, that is to say, is the presence of the void (Lacan *Seminar XI* 174–186, Žižek "Melancholy and the Act"). We did not "lose" nature with capitalism, or resource extraction, or urbanization. Rather, as Lacan puts it, the lost object qua void or lack is what establishes us as subjects. It isn't that "nature," whether by that signifier we mean untouched, primeval forests or a city park, or the historical record that Raymond Williams discusses in his *Keywords* essay, is lost, or is the lost object. Rather, our psychic investment in nature as lost opens up that void which constitutes us as subjects. As Ruti puts it:

> [W]e give up primordial jouissance for the signifier, unmediated pleasure for the capacity to desire. This dynamic becomes crystallized around the fantasy of having lost the Thing, the original (non) object that (supposedly) offered unmitigated jouissance. Nothing of course was lost in reality; there never was any unmitigated jouissance. But the fantasy of having lost this jouissance brings us into being as subjects of lack who experience ourselves as having been deprived of something unfathomably precious.

(*D* 108)

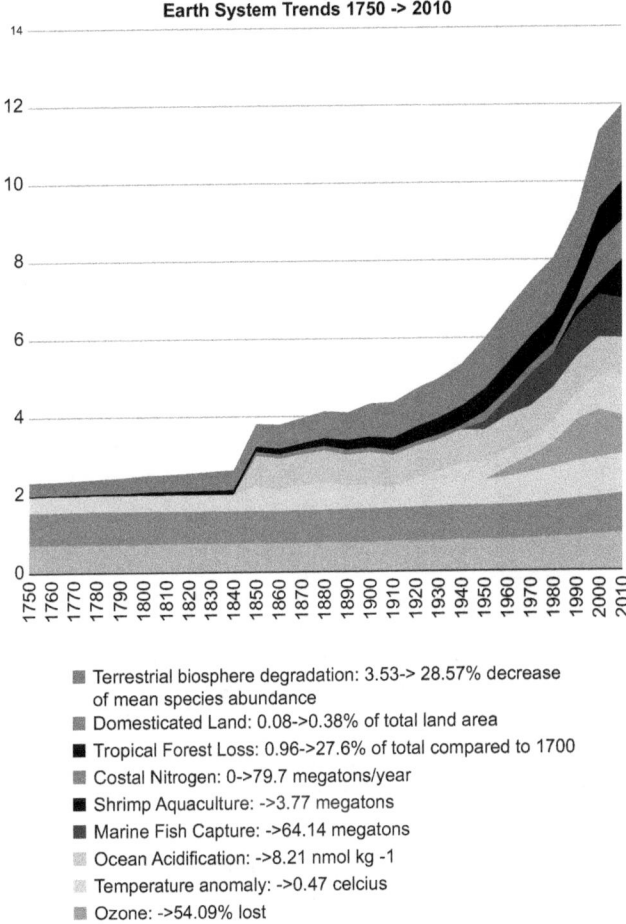

Earth System Trends 1750 -> 2010

■ Terrestrial biosphere degradation: 3.53-> 28.57% decrease
   of mean species abundance
■ Domesticated Land: 0.08->0.38% of total land area
■ Tropical Forest Loss: 0.96->27.6% of total compared to 1700
■ Costal Nitrogen: 0->79.7 megatons/year
■ Shrimp Aquaculture: ->3.77 megatons
■ Marine Fish Capture: ->64.14 megatons
▩ Ocean Acidification: ->8.21 nmol kg -1
▩ Temperature anomaly: ->0.47 celcius
■ Ozone: ->54.09% lost
▩ Methane: 705.34->1744.07 PPB
■ Nitrous Oxide: 271.39->322.46 PPB
▩ Carbon Dioxide: 276.81->384.27 PPM

Figure 4.2 Graph of Great Acceleration (earth system trends) (Bryan
           MacKinnon: usage under Creative Commons Attribution-
           Share Alike 4.0 International license)[3]

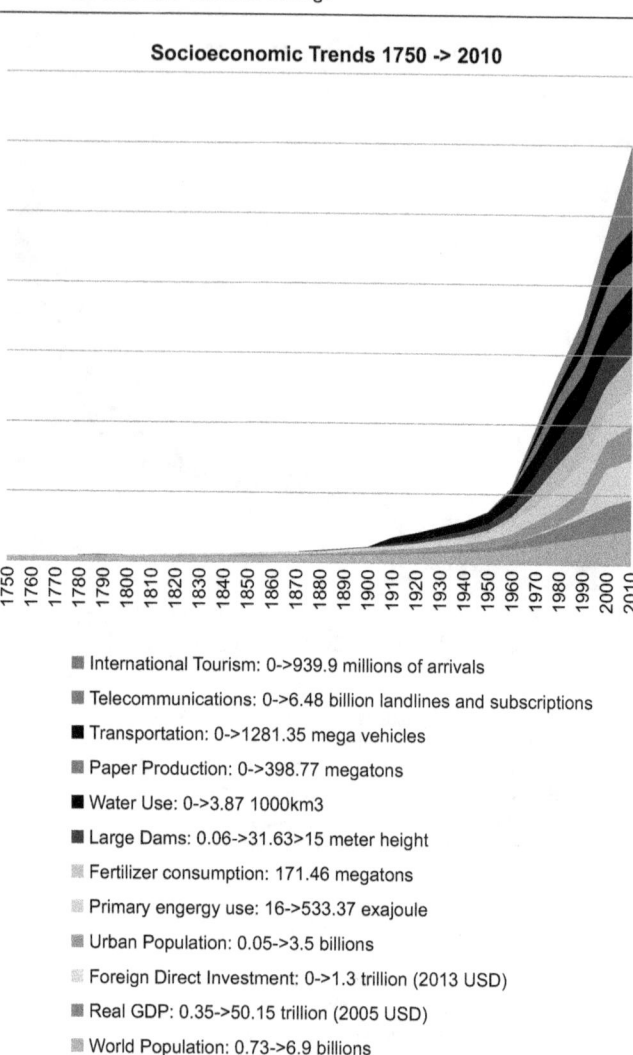

Socioeconomic Trends 1750 -> 2010

■ International Tourism: 0->939.9 millions of arrivals

■ Telecommunications: 0->6.48 billion landlines and subscriptions

■ Transportation: 0->1281.35 mega vehicles

■ Paper Production: 0->398.77 megatons

■ Water Use: 0->3.87 1000km3

■ Large Dams: 0.06->31.63>15 meter height

■ Fertilizer consumption: 171.46 megatons

■ Primary engergy use: 16->533.37 exajoule

■ Urban Population: 0.05->3.5 billions

■ Foreign Direct Investment: 0->1.3 trillion (2013 USD)

■ Real GDP: 0.35->50.15 trillion (2005 USD)

■ World Population: 0.73->6.9 billions

Figure 4.3 Graph of Great Acceleration (socioeconomic trends) (Bryan
MacKinnon: usage under Creative Commons Attribution-
Share Alike 4.0 International license)[4]

We can see an empirical, as it were, verification of this Lacanian thesis with the panoply of data points in what are known as the graphs of acceleration. (I realize I am using one set of diagrams as a way to explicate another diagram, but stick with me. It's worth it, I guarantee.)

Now, there are any number of reasons to be suspicious of these "hockey stick" images, some of which are laid out by Sasha Langford in "The Psychotopology of Climate" (suggestions of linear temporality, *Nachträglichkeit*, and causality), but also their seductiveness as colorful workings of the imaginary, in terms of the specious nature of some of their data, and arguably how they function as phallic signifiers/dick pics. The two graphs of acceleration display first, earth systems, and then socioeconomic trends. The earth systems graph tracks carbon dioxide, nitrous oxide, methane, stratospheric ozone, surface temperature, ocean acidification, marine fish capture, shrimp aquaculture, nitrogen to coastal zone, tropical forest loss, domesticated land, and terrestrial biosphere degradation. The socioeconomic graph tracks population, real GDP, foreign direct investment, urban population, primary energy use, fertilizer consumption, large dams, water use, paper production, transportation, telecommunications, international tourism, and technology. In terms of the second list of data points, the graphs' originators tell us that

> The original 12 included indicators for population, economic growth, resource use, urbanisation, globalisation, transport and communication. We have retained 11 of the original 12 graphs. The only change was to remove the number of McDonald's restaurants, which we used as an indicator for globalisation, and replace it with primary energy use. The combination of foreign direct investment, international tourism and telecommunication gives some sense of the rapidly increasing degree of globalisation and connectivity.
>
> (Steffen et al. 83)

Arguably the swapping out of the McDonald's data illustrates a problem with the project as a whole, akin to Thomas Friedman's "Golden Arches theory of conflict prevention" (which held that "No two countries that both have a McDonald's have ever fought a war against each other").[5] While the latter theory may be put down to journalists' needs to craft snappy explanations, the fungibility of McDonald's, Dell, or primary energy use illustrates the ideological nature of these scientific explanations.

But to return to nature as lost object, there is a way to interpret "earth systems" data points as essentially making the case for loss as constitutive for the contemporary subject. Debates over the term "Anthropocene" miss the point: the issue is not over whether to substitute more "accurate" or politically radical/salient terms like the Capitalocene or Plantationocene. Rather, while the correlation of socioeconomic trends and the earth systems makes a strong case for causation (that is, for the Anthropocene as a different period than the Holocene), the various data points not only exist, especially post 1950, "well outside of the Holocene envelope of variability" (Steffen et al. 92) but also track both habitation loss and/or a leveling off of, say, the growth of agricultural landmass or marine fishing that are due not to a decline in human activity but, respectively, to the end of available land and to the depletion of marine species. Finally, the "Anthropocene" as periodization works in both ways: the decline of the earth systems (nature as loss) is both caused by human activity (however unevenly distributed)[6] and a cause of human subjectivity: the subject of grief.

Beautiful Soul and Bartleby

Still discussing the semiotic rectangle of climate grief, what of the left side, the non-grieving subject and their relation to nature? Here the dyad of Hegel's Beautiful Soul and Melville's Bartleby are helpful. The non-grieving artist is not, I assert, the Beautiful Soul, but also the activists do not, typically, grieve. Activists are quite impatient with the discourse of climate grief, believing, not unreasonably, that this is the narcissistic affect of those who are suddenly, belatedly, "aware" that the world is out of joint.[7] But the petro-prole, the carbon subject, does grieve, most decidedly, and in many ways theirs is a version of the Bartlebyan "I would prefer not to," which last position is also, as Alma Krillic has argued, implicated in the Kübler-Ross stages of grief.

The Hegelian Beautiful Soul is trapped in an antinomy where, on the one hand, its morality is dependent on not acting: "It lives in dread of besmirching the splendor of its inner being by action and an existence, and, in order to preserve the purity of its heart, it flees from contact with the actual world" (Hegel, *Phenomenology*, §658, pg. 400).[8] But on the other hand, that morality has to be known, to be communicated: "On account of this utterance in which the self is expressed and acknowledged as an essential being, the validity of the act is acknowledged by others" (*Phenomenology*, §656, pg. 398). The figure is a result of trying to square the Kantian circle of morality (or duty) and freedom, with the finger on the scale of the latter by disavowing its necessity:

> It renounces all these attitudes and dissemblances, connected with the moral view of the world, when it renounces that consciousness which thinks of duty and reality as contradictory. According to this latter view, I act morally when I am conscious of performing only pure duty and nothing else but that; this means, in fact, when I do not act. But when I really act, I am conscious of an "other", of a reality which is already in existence, and or a reality I wish to produce; I have a specific purpose and fulfil a specific duty in which there is something else than the pure duty which alone should be intended.
>
> (*Phenomenology*, §637, pg. 386)

The problem for Hegel is that the only consciousness or thought possible being self-consciousness, once one is aware one is acting, the awareness of that takes over. It isn't that the Beautiful Soul doesn't do anything or wants to keep its hands clean. It's rather that it is hoisted on its own petard: knowing I am acting means I am not acting, I am knowing. Make no mistake, the Beautiful Soul is no cardboard cutout, no risible virtue-signaler. Rather, its formulation in Hegel (the original is perhaps Jesus Christ, but Hegel's more proximate models are the heroes of German Romantic literature, in Goethe, Jacobi, and Schiller) is one of true contemplation, and the value thereof. Jean Hyppolite conveys the contradictions well after spending some pages on the figure in his Hegel book: "The beautiful soul represents the universality of spirit which is certain of itself but which is determinate because it opposes the partiality of action" (517).

Two important discussions of the Beautiful Soul are missed encounters, if you will, rather as that of the Beautiful Soul and Bartleby. In the first, Žižek accuses Lacan of mixing up Hegel's Beautiful Soul with another "denizen" of Hegel's zoo (Pippin), the Law of the Heart; in the second, Rei Terada argues that Rebecca Comay lets Hegel off the hook (or even eggs him on) in his dismissal of the Beautiful Soul who, in Terada's *soi-disant* "afropessimist" reading, is a Social Justice Warrior. Here is Žižek:

> In his reference to the Hegelian Beautiful Soul, Lacan makes a deeply significant mistake by condensing two different "figures of consciousness": he speaks of the Beautiful Soul who, in the name of

her Law of the Heart, rebels against the injustices of the world. With Hegel, however, the "Beautiful Soul" and the "Law of the Heart" are two quite distinct figures: the first designates the hysterical attitude of deploring the wicked ways of the world while actively participating in their reproduction (Lacan is quite justified to apply it to Dora, Freud's exemplary case of hysteria); the "Law of the Heart and the Frenzy of Self-Conceit," on the other hand, clearly refer to a psychotic attitude – to a self-proclaimed Savior who imagines his inner Law to be the Law of everybody and is therefore compelled, in order to explain why the "world" (his social environs) does not follow his precepts, to resort to paranoiac constructions, to some plot of dark forces (like the Enlightened rebel who blames the reactionary clergy's propagating of superstitions for the failure of his efforts to win the support of the people). Lacan's slip is all the more mysterious for the fact that this difference between Beautiful Soul and the Law of the Heart can be perfectly formulated by means of the categories elaborated by Lacan himself: the hysterical Beautiful Soul clearly locates itself within the big Other, and it functions as a demand to the Other within an intersubjective field, whereas the psychotic clinging to the Law of one's Heart involves precisely a rejection, a suspension, of what Hegel referred to as the "spiritual substance."

(Žižek, *Tarrying* 267n)

Now, I don't know about you, but the Beautiful Soul and the Law of the Heart sound pretty similar, and what they look like is precisely the activist or climate organizer, the Greta Thunberg, Wangari Maathai, or the Tiny House Warriors.[9] Or are they just a caricature of same? Here is how Rei Terada puts it, when she argues that Hegel's concept of the Beautiful Soul is politically suspect:

According to Hegel the Beautiful Soul, as a personification of "morality," denounces society, like a Social Justice Warrior of "privatized ethics," and "does not possess the power to renounce the knowledge of itself which it keeps to itself," [Hegel §668] knowledge of its own complicity . . . [i]nsofar as it lacks the power of double negativity.

(Terada 141)

But the problem here is that due to Terada's own identification with the Beautiful Soul qua Social Justice Warrior, not only does she read Comay's account of Hegel's figure as merely a nasty swipe at the mad or the ill (Hölderlin or Novalis), but her over-hasty and anachronistic labeling of Hegel's concept of the Beautiful Soul as akin to the Social Justice Warrior depends itself on a misreading of Drew Milne's notion of "privatized ethics," discussed further later, and also an anti-intellectual notion of Hegelian dialectics.

Terada continues:

> The Beautiful Soul's demise in self-consumption (Novalis died of tuberculosis at the age of twenty-eight; Hölderlin suffered from a debilitating mental illness), and the critical response to it, shows these limited possibilities. [Drew] Milne points out that Hegel's allusion to Novalis's tuberculosis becomes an occasion for wider moralization, as critics take the opportunity to either call the allusion cruel or imply that the Beautiful Soul deserves it. Or both. Comay mentions that the "biographical signposts . . . cruelly, ludicrously" make fun of Novalis's death. Then, praising the precision of the metaphor, she joins in: "the infinite contraction of a subjectivity withdrawn into itself (as Hegel later characterizes madness) presents the mirror image of the infinite dispersal of a body reduced to its elemental atomic particles so as to disappear without a trace."
>
> (Comay 116; ctd. Terada 143)

First then, in *Mourning Sickness*, Comay writes the following:

> As the beautiful soul becomes "conscious of its contradiction" (to wit, the antinomy between the universality to which it is devoted and the fanatic singularity of its devotion), it becomes "disordered to the point of madness, wastes away in yearning and pines away in consumption" (PhG §668). Hyperbole aside, the madness has a specific historical inflection. . . . Whatever the conceptual link between "madness" and "consumption" – between Hölderlin babbling in his tower and Novalis at his tubercular last gasp (for in these two directions the biographical signposts seem all irrefutably, cruelly, ludicrously to point) – the association is interesting not least because in both cases we are presented with an immediate convertibility of psyche and soma.
>
> (Comey 116)

My reading of this passage in Comay differs from Terada insofar as rather than "praising the precision of the metaphor" (Terada 143), I see

Comay as arguing that the metaphor is first historical, and then, in its modernity, pointing to its internalization of morality. (But Comay is also not given to the position of the grieving subject.) This last point is akin to what Milne refers to as "privatized ethics" (Milne 65, ctd. Terada 141), but here again an expansion of the source material is helpful:

> Capitalism's intensification of secular conflicts with religious dogmatism opens this chasm to one of global proportions. In this sense, the beautiful soul can be understood as a symptom of secular modernity, the subject of privatized ethics that seek to separate moral thought from the aporia configured and reconfigured in political and legal institutions. Shorn of credible frameworks or institutions of duty, the beautiful soul lives through the aspiration to have an inner beauty of moral feeling without recognizing heteronomous authorities. Attempts to conceptualize such aspirations motivate postmodern ethics and utopian forms of cosmopolitanism implicit in hopes for multicultural tolerance. A distant relative of the beautiful soul can be discerned in the paralysis experienced by someone newly sensitized to the moral claims of the Other, where otherness is conceived of as infinite respect for the radical incompatibility of competing cultural, moral or legal conceptions of the person.
>
> (Milne 65)

Milne maintains the notion of the Beautiful Soul as a figure of modernity, one who, in Lacanian or Žižekian terms, confronts the demise of the big Other ("Shorn of credible frameworks or institutions of duty, the beautiful soul lives through the aspiration to have an inner beauty of moral feeling without recognizing heteronomous authorities" [Milne 65]). Just to cut to the chase: if grief, as the Marianne Jean-Baptiste epigraph to this book suggests, is often narcissistic ("There's nothing rational about grief – maybe you're crying for yourself") and not rational, perhaps Terada's special pleading for the Social Justice Warrior (her own fantasy via Comay via Hegel – perhaps I have the order wrong, I do not know) is similar. Rei, you're crying for yourself.

The Beautiful Soul then becomes, in Melville's hands, Bartleby, a scrivener or legal copyist who tells his overseer, "I would prefer not to." But the story is narrated by that boss, who runs down his list of workers, all with Dickensian nicknames – Turkey, Nipper, Ginger-Nut. Turkey, intriguingly, seems to be a proto-carbon subject: we are told "his face was of a fine florid hue, but after twelve o'clock, meridian – his dinner hour – it blazed like a grate full of Christmas coals" and later in the day he made all kinds of mistakes – blots his copybook – and "his face flamed with augmented blazonry, as if cannel coal had been heaped on anthracite" (Melville 61, 62). As if to presage in reverse the later staff problems, the lawyer boss would tell Turkey to go home on Saturday afternoons, but he would refuse. Bartleby is a new hire, and at first is a swift and competent copyist, but by day three, when asked to help the lawyer check some work (proofreading being part of the drudgery), he first utters the famous "I would prefer not to" (Melville 68). This is said on another proofreading day when the entire office is proofreading a text, in this case copies that Bartleby had made. The narrator is stymied by what he calls "passive resistance" (Melville 72), and as the story progresses, Bartleby prefers not to do more and more, including leaving the office, until he is arrested and ends his days in the local prison, where he dies. But, in a twist on Hegel's conundrum (knowing that I act, I do not act), Bartleby perhaps does not know: "Poor fellow! thought I, he means no mischief; it is plain he intends no insolence; his aspect sufficiently evinces that his eccentricities are involuntary." Is "preferring not to" Bartleby's desire or the desire of the other, is it an unconscious wish?

Melville's figure is one of the most well-raked over literary characters of Ruti's "progressive critical theory" – rather like the Christmas coals to which Turkey's complexion is compared. "Bifo" (Franco Berardi) finds in Bartleby a figure for what he calls "passivism" (Berardi 128); for Jodi Dean, channeling Hardt and Negri, Bartleby is "a figure of refusal, opposition, or resistance, a model of escape or disentanglement" but somehow evades the post 9/11 protestor's enjoyment (Dean 68); while the opposite is true for Joan Copjec, ventriloquizing Agamben: "Bartleby becomes the exemplary figure of . . . impotentiality, the first manifestation of a subject's power or capacity. Psychoanalysis, we well know, names this capacity libido (or *jouissance*)" (Copjec 171), and if Žižek finds "Bartleby does not negate the predicate; rather, he affirms a non-predicate: he does not say that he *doesn't want to do it*; he says that he *prefers (wants) not to do it*" (Žižek *Parallax* 381), Deleuze is fixated on the agrammaticality of the proposition. C.L.R. James, however, in two-and-a-half pages (like much of the book in which it appears, "written on Ellis Island while I was being detained by the Department of Immigration") provides a summary of "Bartleby" that, for all its admiration, displays also a trenchant analysis of its usefulness to the Marxist of the 20th century:

> The story also explains Melville's life-long distrust and scepticism of liberals, radicals and revolutionaries, and his permanent disrespect for the Congress of the United States. What did all these people have to say, what could they do about a life such as Bartleby's? His answer was: nothing.
>
> (James 115)

In her essay on Bartleby and environmentalism, Krillic argues that while one could read Bartleby, *à la* "Bifo," as simple passivity in the light of climate disasterism (she cites, as well, Donna Orange), she insteads view his (in)action as a matter of enjoying his environment and, taking a page from Deleuze's book, argues that the very grammatical structure of his utterance "I would prefer not to" bespeaks a lack of motivation: "no driving force behind Bartleby's words, no apparent desire to change the existing order of power, and no motivation to outright resist his employer" (Krillic 64). Byung-Chul Han, however, notes that the very environment of the story bespeaks a disciplinary ethos: "Walls and partitions, the elements of disciplinary architecture, traverse the entire narrative" (Han 26). But there is a way to see instead Han's analysis ("a lack of drive and . . . apathy, which seal Bartleby's doom" [Han 26]) as one of a society of control: "Bartleby works behind a screen and stares empty-headedly" (Han 26). Of course, Bartleby's screen is not a computer screen, nor even the fabric covered cubicles Douglas Coupland called "veal-fattening pens" in his *Generation X* novel, but Han willfully misreads the story's subtitle "A Story of Wall-Street" as more evidence of that disciplinary architecture. Need I remind you we are here for the Kübler-Ross? Alma Krillic's genius move in her essay is to ascribe the stages of grief or mourning not to Bartleby (who, after all is just stuck) but to his boss. Who speaks up for the boss, after all? Well, pretty much everybody, or at least everybody whose voices are heard in the corridors of power, as opposed to the hoi-polloi of social media and press scrums. "I believe that the lawyer's reactions to Bartleby's preference not to work resemble the five stages of mourning" Krillic tells us, and that "[t]he loss that the lawyer was experiencing was the loss of power over the situation and he thus moves through the stages of grieving his loss of control" (Krillic 62). So, we have passivity ("Bifo") and not (Krillic), *jouissance* (Krillic and Copjec) and not (Dean), caught up in the pedantry of grammar (Žižek and Deleuze), disciplinary society (Han) and not (Han), and rather than an example of Kübler-Ross instead its instigation (Krillic).

What does it mean to say that the petro-prole, the carbon subject, is a Bartlebyan subject? What do they prefer not to do, is this a form of denial or disavowal? Certainly, Žižek's elaboration makes space for that refusal of politics:

> We can imagine the varieties of such a gesture in today's public space . . . "Are you aware how our environment is endangered? Do something for ecology!" – "I would prefer not to." This is a gesture of subtraction at its purest, the reduction of all qualitative differences to a purely formal minimal difference.
>
> (*Parallax* 384)

As Hern and Johal's interviewees made clear in the previous chapter, it isn't as if the petro-proles do not know they're participating in making the planet worse.

Also, why is Bartleby, as we have seen, held up as much more of a role model by the "progressive critical theory" than the Beautiful Soul, where the latter seems to apply more to theory/theorists themselves, and, frankly, Bartleby seems to depend, for all Dean's and Žižek's protesting (too much), on the big Other of the office job? Is the Beautiful Soul all that bad, or is it perhaps more radical than Bartleby because of the value of self-consciousness? As stated, "all that now remains to be done is to supersede this mere form, or rather, since this belongs to *consciousness as such*, its truth must already have yielded itself in the shape of consciousness" (Hegel, *Phenomenology*, §788, pg. 479). Self-consciousness had to be the object of consciousness for absolute knowing. Hyppolite describes Hegelian self-consciousness in a way that kinda sounds like COVID-19 brain fog, consciousness of one's knowledge rather than consciousness of the object. Because you can't remember or think of what it is you're trying to think of, you're just aware that you are thinking. Perhaps the victim of climate change, who cannot leave the house because of forest fire smoke (no going for a run, asthmatics and the elderly and children should also stay inside) is like the Beautiful Soul/Absolute Knowledge, because what else can you do but stay inside and read some "progressive critical theory"? Learn, learn, learn!

And yet, "progressive critical theory" is, after all, Ruti's term in *The Ethics of Opting Out*. As happens so often for theory books' titles, just what that opting out entails is left undefined in *EOO*, or, perhaps, that's what you have to read the book to find out. Is opting out a matter of queers not getting married or of queers opting out of late-stage capitalism? That is, is opting out a matter of the Beautiful Soul, so just keeping one's hands clean, or is it akin to the Bartleby position, a more principled refusal?

Hockey sticks and kettle logic

To develop this argument, and by way of concluding, I want to think about the "coincidence of opposites" of climate denialism and activist science via the tropes of Freud's kettle logic. As everybody knows, the first dream Freud analyzes in *Interpretation of Dreams* is one of his own, the dream of Irma's injection. Toward the end of his analysis, he admits that many of the dream's elements contradicted each other (it wasn't his fault his patient didn't listen to him, or her suffering was due to bodily, and not psychic, causes, or was because she was a widow, or because a colleague had given her an injection of a problematic drug, or the needle was dirty):

> The whole plea – for the dream was nothing else – reminded one vividly of the defence put forward by the man who was charged by one of his neighbours with having given him back a borrowed kettle in a damaged condition. The defendant asserted first, that he had given it back undamaged; secondly, that the kettle had a hole in it when he borrowed it; and thirdly, that he had never borrowed a kettle from his neighbour at all. So much the better: if only a single one of these three lines of defence were to be accepted as valid, the man would have to be acquitted.
>
> (Freud, *Interpretation* 119–120)

In Joseph Dodds's *Psychoanalysis and Ecology at the Edge of Chaos*, Freud's theory of "kettle logic" is applied to various defenses mobilized against the anxiety of climate change: "there's nothing wrong with the climate/kettle," "there was a hole in the planet when you gave it to me," and "there's nothing we can do about it" (Dodds 43–44). "This joke is particularly useful in our context," Dodds continues,

> because of its structure (mutual contradictions united by a common motivation), the fact that a kettle is, like our climate, a container that can be broken when heated up beyond a certain limit,

and because the joke's formula corresponds well to the many arguments against action on climate change (often argued by the same
person simultaneously).

(Dodds 43)

What is important here, I am arguing, is how the structure of climate
denialism mimics that of Freud's insights into the dreamwork or the
jokework,[10] which is to say, the unconscious. Žižek drew on the "borrowed kettle" for his analysis of the (il)logic justifying the 2003 invasion of Iraq: Saddam Hussein possessed weapons of mass destruction,
but then they couldn't be found, but then even if there is no connection
between Iraq and 9/11 he's a murderous dictator, and so on.[11]

A similar homology can be found in Michael E. Mann's *The
Hockey Stick and the Climate Wars*, where he declares that there is "a
hierarchy to the denialist canon" that ranges from "CO_2 is not actually
increasing" to "Even if there is warming, it is due to natural causes,"
all the way to "Whether or not the changes are going to be good for us,
humans are very adept at adapting to changes; besides, it's too late to
do anything about it, and/or a technological fix is bound to come along
when we really need it" (Mann *Hockey Stick* 23: see also Buckley and
Szapudi). In both Dodds's and Mann's cases, their logic is impervious
to contradiction, suggesting what we might call a "climate denialist
unconscious." Climate scientists hew strongly to the university discourse: rationalist, quantitative, shorn of jouissance or ideology. This
is clear in Mann's discussion of the role that tree ring data plays in
the 2009 "Climategate" scandal,[12] where denialists seized on a 1999
email that referred to a "trick" used to swap out so-called proxy data
(tree rings, coral, ice samples) and instrumental temperature records,
among other charges denied by Mann in a *Wall Street Journal* op-ed
where he argued that "men and women who have dedicated their lives
to advancing science need not apologize for keeping their rigorous
professional journals free of the pollution of what is purely politics"
(Mann "Science Journals").

Mann's defense of peer review and the authority (sanctity?) of scientific journals falls into the same category error for which he accuses
the more tendentious climate denialists: argument in the public sphere
and not, for instance, that of science studies or rhetorical or print culture.[13] But Mann's recent victory over the denialists in the courts may
help us understand the Lacanian, if not Ruti-esque, valences of this debate. As reported in *The Washington Post*, the right-wing blogger Rand
Simberg and *National Review* writer Mark Steyn were ordered to pay

more than $1 million in a defamation lawsuit. Simberg wrote, in a 2012 column on the website of the Competitive Enterprise Institute, that

> Mann had "molested and tortured data" of global warming and compared Mann to [Jerry] Sandusky, who was a Pennsylvania State University football coach who had been arrested for molesting young boys At the time, Mann was a professor at Penn State. In the *National Review*, Steyn quoted the article and added: "Not sure I'd have extended that metaphor all the way into the locker-room showers with quite the zeal Mr. Simberg does, but he has a point."
>
> (Grandoni)

Jacques-Alain Miller tells us that Lacan, in *Seminar III*, argued that "from the Freudian point of view man is the subject captured and tortured by language." Mann, as Paul Kingsbury quipped, sued his detractors for calling him a Lacanian.[14]

Further, we can see Mann's hierarchy of "the denialist canon" (which almost sounds like a fanfic utterance) something close, if not to Kübler-Ross, to Žižek's appropriation in *Living in the End Times*, where the five stages of denial, anger, bargaining, depression, and acceptance are, respectively, aligned with the analysis of ideological obfuscation, violent protests, political economy, subjective pathologies, and the emergence of the ethical-revolutionary subject (Žižek 2010). This concatenation of kettle logic and Kübler-Ross, silver ring and *das Ding*, tree rings in Hitchcock and Mann, cli-fi and autotheory, then, allows for some preliminary conclusions on what Mari Ruti's theory teaches us about our attitude toward climate change.

I have two propositions. In an important paper on sublimation, Krzych shows how for Ruti, "[w]hen the world seems to fall apart . . . when we *fall from fantasy* – the experience of such trauma may entail for a subject new opportunities for creative expression" – this is to see grief, and the object providing us access to a new relation to our singularity (remember, this is idiosyncratic, contingent); but then I also said I have some problems with the notion that the ontological lack (which any card-carrying Lacanian sees as constitutive due to our being speaking subjects) can provide us with the "creative, innovative capacity to endure the kinds of more contingent lacks" (Krzych). As Ruti writes:

> From a Lacanian viewpoint it is only the fall of our most treasured fantasies – particularly of the idea that there is some "sovereign

good" that is capable of shielding us from the terror of living – that allows us to transition to a more imaginative and creatively engaged psychic economy. More specifically, the disbanding of fantasies enables us to better listen to the idiosyncratic particularity of our desire, and in so doing to begin to forge a singular identity apart from the social conventions that seek to determine the parameters of our being.

("Fall of Fantasies" 486)

Derek Hook discusses these "libidinal treasures" (a cognate of Ruti's "treasured fantasies") in his analysis of the "theft of enjoyment" as mechanism for racism and anti-Semitism (Hook 42).

A similar dynamic is at work in today's *coincidentia oppositorum*, where the treasured fantasies of the left have been stolen by the anti-vaxxer and populist right, with antecedents in the corporate wing of the climate denialist movement. Consider Naomi Klein, regarding her book *Doppelganger*, in which she discusses not only the online confusion of herself and the right-wing Naomi Wolf, but also the post-COVID-19 horseshoe convergence of left and right:

This pattern is rampant: on the one hand, *appropriation and trivializations of precious language and symbols* used to fight real state terror and cover-ups – on the other hand, aggressive attacks on those very movements, including in the pro-cop renaissance, the abortion bans, the wave of book banning and other attempts to make knowing true history an illegal act.

(Klein 2023, n.p.; my emphasis)

That is, without venturing too far into the realm of Lacanian centrism, perhaps Ruti and Hook's analysis can be turned around on Klein's melancholy. Here I am working on a *détournement* of the radical's reminder that there is a fascist inside all of us – perhaps there is a communist inside the Freedom Convoy.

Naomi Klein's assemblage of rightist actions reminds us of how, in today's political climate, various attempts are now being made to conceptualize the layering or stacking of climate change, the pandemic, populism, and state anti-Black violence: from the more anodyne "polycrisis" to the more activist "intersectionality" (both partaking, I would argue, of the university discourse) and my own term "clusterfuck," which I prefer for its libidinal violence. Nonetheless, it is important to recognize that in much of the West, the opposition to trans rights, or vaccine mandates, overlaps not only with climate denialism but is part of the 21st-century growth in various forms of folk skepticism toward science alongside the rise of conspiratorial thinking and populism (from anti-vaxxers and climate denialism, and the invasion of Ukraine to electoral malfeasance, paranormal discourses, and opposition to trans medical access) along three vectors: an account of the "polycrisis" of events and conditions ranging from COVID-19 and anti-Black violence to climate change (Burnham, "*Nil actum credens*," Žižek, *Freedom*); the question of whether to see such skepticism or denialism as the effects of top-down institutional misinformation, or inherent in liberal-democratic politics; and a Lacanian theory of the subject and knowledge grounded in a "passion for ignorance" and the ethical act.

As I mentioned at the start of this book, I am thinking about Ruti and climate grief in the context, as well, of the recent death of my father. In an odd, if not contingent fashion, Freud very early on considered paranoid affect to be one related to the death of a parent. An editorial note to "Mourning and Melancholia" points us to

> a manuscript, . . . addressed to Fliess, and bearing the title "Notes (III)" . . . dated May 31, 1897. . . . The passage in question, whose meaning is so condensed as to be in places obscure, deserves to be quoted in full.
>
> (Freud "Mourning" 240)

Freud's note:

> Hostile impulses against parents (a wish that they should die) are also an integral constituent of neuroses. They come to light consciously as obsessional ideas. In paranoia what is worst in delusions of persecution (pathological distrust of rulers and monarchs) corresponds to these impulses. They are repressed at times when compassion for the parents is active-at times of their illness or death. On such occasions it is a manifestation of mourning to reproach oneself for their death (what is known as melancholia) or to punish oneself in a hysterical fashion (through the medium of the idea of retribution) with the same states [of illness] that they have had. The identification which occurs here is, as we can see, nothing other than a mode of thinking and does not relieve us of the necessity for looking for the motive.
>
> (Freud "Mourning" 254–255)

And so it is also possible to see the climate denialist as an example of the ethical subject, who, committed to the singularity of their desire, does not give up on their jouissance, sees that there is no big Other, a radicality of subjectivity that should teach us something about our own political position. Climate deniers' (or anti-vaxxers, etc.) unconscious, and the fungibility of grief, means we have to take Lacan's nostrum in *Seminar VII*, that the only thing you can be guilty of is ceding your desire: the climate denier is the ethical subject per Lacan. Like Antigone burying her brother against the law of the state, they do not cede their desire. This proposition, as I have learned while presenting material from this book in the past two years, is evidently contentious. But I think if Lacanians only find ethical subjects whose politics agree with ours, we are remaining in the imaginary. There is a communist in us all. Fredric Jameson has said that even the most odious of ideologies – all the way to Naziism – contain a kernel of the utopian. Thinking of the Freedom Convoy or the January Sixer or the climate denier as ethical subject can help us both understand what a Lacanian ethics really consists of, and what our desire is, in the apocalyptic, but also anxious, times.

We can only grasp the radical form of Lacan's ethics if we admit it leads in directions we aren't comfortable with. I do not think politics is the last horizon. Rather, I'm attempting a kind of bench test of the concept of the ethical subject, an opening to analysis, which I think is beyond political good and evil. Also, if the so-called ethical subject is simply not ceding their desire in terms of goals we agree with, that's just our desire, the big Other. Rather, we have to ask, in what way do they touch the Real? This means to (re)turn to Antigone. On the one hand, if it's Lacanian desire, there has to be an internal vortex of doubt, negation, lostness. Loss of loss, question of lack in the subject versus lack in the big Other, dialectic of desire – not ceding one's own desire, but desire is the desire of the Other (that is, desire is that antagonism). Is Antigone an ethical subject? I think she is, and that Lacan says she is. She does not cede her desire – to bury Polynices. Regardless, it's important that Lacan also says in *Seminar VII* that we don't have to have emotions at the Greek tragedy – we can be distracted thinking of our pen or the check we were meant to sign – that is to see tragedy in terms of the chorus has emotions for us, it's not cathartic, it's not identifying with Antigone, who, on the contrary, refuses the sheeple's demand, puts everything on the line, goes beyond the limit, the *Atē*. Lacan is all about edginess – change, disruption – especially against the so-called revolutionary 1968 students who wanted change and got neoliberal universities.

Lacanian centrism and centrist Lacanianism

I said earlier that we should be wary of venturing into Lacanian centrism. Ruti's work forces us to face such a proposition, first, that there is such a thing as Lacanian centrism, which is not the same thing as centrist Lacanians. First, all or most Lacanians are centrists, by which I mean the middle-class professoriat and clinical professionals, whose politics, whether electoral or having to do with gender and other social movements, are for the most part liberal but also accommodationist. Lacanians may profess revolutionary theory, read Marx or Hegel, espouse anti-patriarchal views in the classroom and on the pages of their books, but they own houses, get married, have pensions, and the rest of it. They are good citizens. They vote and are aghast at populist "acting out." If the populist canard is that academics are all Marxist lunatics, well, that's as much a misrecognition as it is satisfying to the professoriat; thus, Lacanian centrists are professors who happen to read Lacan. In this regard, Ruti's critique of Žižek and Badiou in terms of gender is more radical. She can be quite trenchant:

> No matter how genuinely "universalist" the intentions of Badiou and Žižek may be, their neo-Marxist theories repeat the masculinist and white-hegemonic weaknesses of classical Marxism so that while class (or one's status as a member of the "proletariat") qualifies as a "universal" basis for progressive struggle, race, ethnicity, gender, and sexuality do not. As a matter of fact, these thinkers are not content to merely exclude movements that target racism, ethnocentrism, sexism, and homophobia, but actively disparage them.

> (*SB* 204)

Indeed, Ruti is able to turn Badiou and Žižek's propensity for generalized universalism (and anti-identity politics) against them, arguing that queer, or postcolonial, or anti-racist subjects lack substance every bit as much as the proletariat, and that Badiou and Žižek's "universalism does not even begin to meet the requirements of a genuinely universalist universalism" (*SB* 215).

If Ruti is to the left of the Lacanian centrists, her position, in terms of Lacanian politics, is itself centrist, what I am calling centrist Lacanianism. Within Lacanian theory proper, the politics of drive and the ethical act, the sinthome, constitute an extension of that theory beyond the bounds of the state or the social. It is a politics of dissolution, as argued in the late Lacan, his *L'Étourdit*, and his 1980 act of dissolving his school, in Edelman, in the Badiou of the *Theory of the Subject* and his *Lacan* seminar. In regard to this position, Ruti is a centrist.

This is the importance of her work for helping us to theorize climate grief: as I said in Chapter 2, the question of the fungibility of grief, of whether grief stacks, or we deal with one then another, or they make each other worse, is the most important contemporary politics of grief. Add to this Ruti's arguments (for they are not one monolithic argument, especially when "BB" and "WCNC" are read alongside her earlier work, alongside *EOO* in particular) that, first, there are different kinds of lack, the constitutive, the contingent, and the socioeconomic, and, then, that the sublimation one works up in terms of Lacanian lack helps one deal with the contingent lack, or the confrontation with socioeconomic lack forces one to confront one's constitutive or Lacanian lack. Ruti's contradictory position, as more radical than Lacanian centrists, but herself a centrist Lacanians, then, is key to understanding all of this, for which we now turn to this book's coda.

Notes

1 Y. Agematus, *ZIP: 01-01-14 . . . 12-31-14*. Vancouver: Artspeak/ New York: Thea Westreich & Ethan Wagner/Portland: Yale Union, 2015.
2 Binners' Project. https://www.binnersproject.org/ See also Porter.
3 https://creativecommons.org/licenses/by-sa/4.0/deed.en
4 https://creativecommons.org/licenses/by-sa/4.0/deed.en
5 Friedman, "Foreign Affairs Big Mac." See also Friedman's "Dell theory of conflict prevention" according to which "The Dell Theory stipulates: No two countries that are both part of a major global supply chain, like Dell's, will ever fight a war against each other as long as they are both part of the same global supply chain" (Friedman, *The World is Flat* 587).
6 A. Malm and A. Hornborg, "The Geology of Mankind? A Critique of the Anthropocene Narrative," *The Anthropocene Review* 1 (2014): 62–69.
7 Here I draw on conversations and presentations at the *Facing Ecological Grief* symposium, SFU Faculty of Environment, Vancouver, April 2023. I am particularly grateful to Rita Wong for her forthright, and brisk, criticisms of climate grief discourse.
8 In the "Confessions of a Beautiful Soul" (*Wilhelm Meister*), Goethe has his heroine express the same thought: "I would willingly

leave my parents and earn my bread in a foreign land rather than act contrary to my thoughts," or again: "In the face of public opinion, my profound conviction and my innocence were my surest guarantees" (Ctd. Hyppolite 501).

9 Wangari Maathai won the 2004 Nobel Peace Prize for her work over the decades fighting deforestation in Africa, leading to the "Green Belt Movement" ("Wangari Maathai"). The Tiny House Warriors are Indigenous land defenders on Canada's West Coast, fighting pipelines in Secwepemc territory ("Tiny House Warriors").

10 Dodds continues: "For the purposes of simplification these can be summarized in three positions, progressively accepting more of the reality of the ecological crisis but all resulting in inaction: 1. It's not happening; 2. It's not my fault; 3. There's nothing we can do about it (so I can just get on with my life as usual)." What fantasies, anxieties and defences, Dodds asks, are expressed by these positions? (Dodds 43).

11 However in the case of today's most pressing wars, Russia's 2022 invasion of Ukraine and then the 2023 Zionist invasion of Gaza, the germane element of Freud's joke is not "kettle logic" but rather the figure of the neighbor: for Russia or Israel the biblical or Torah injunction to "love thy neighbour," as it was for Freud and Lacan, is rather too much, a monstrous proposition.

12 "Climatically useful tree ring proxies are thus primarily restricted to the midlatitude continental regions, leaving much of the globe un-sampled by this method" (Mann, *Hockey Stick* 32).

13 See, for instance, Ramírez-i-Ollé, who argues that such practices constitute "boundary work" insofar as "[s]cientists demarcate the boundaries of science by attributing select characteristics to the knowledge, methods and practitioners that constitute the institution of science in order to achieve professional authority and wealth" (Ramírez-i-Ollé 3).

14 Paul Kingsbury, personal communication.

References

Berardi, F. "B." *Breathing: Chaos and Poetry*. New York: Semiotexte, 2018.

Burnham, Clint. "Lacan's Trash Talk: Three Objects for the Internet." In Burnham and Kingsbury, Eds., 75–93.

Burnham, Clint. "*Nil actum credens, si quid superesset agendum*: Or, Slavoj, Can't You See I'm Burning? Žižek *avec* the Clusterfuck of 2020." *Understanding Žižek, Understanding Modernism*.

Eds. Jeffrey Di Leo and Zahi Zalloua. London: Bloomsbury, 2022. 77–89.

Burnham, Clint and Paul Kingsbury, Eds. *Lacan and the Environment*. London: Palgrave, 2021.

Comay, Rebecca. *Mourning Sickness: Hegel and the French Revolution*. Stanford: Stanford UP, 2010.

Copjec, Joan. "The Object-Gaze: Shame, Hejab, Cinema." *Filozofski Vestnik* (2006): 163–182.

Dean, Jodi. *Blog Theory: Feedback and Capture in the Circuits of Drive*. Cambridge: Polity, 2010.

Deleuze, Gilles. "Bartleby; or, the Formula." *Essays Critical and Clinical*. Trans. Daniel W. Smith and Michael A. Greco. Minneapolis: U of Minnesota P, 1997. 68–90.

Dodds, Joseph. *Psychoanalysis and Ecology at the Edge of Chaos: Complexity Theory, Deleuze, Guattari and Psychoanalysis for a Climate in Crisis*. London: Routledge, 2011.

Edelman, Lee. *No Future: Queer Theory and the Death Drive*. Durham: Duke UP, 2008.

Freud, Sigmund. *The Interpretation of Dreams*. 1900. *The Standard Edition of the Complete Psychological Works of Sigmund Freud*. Volumes 4 and 5. Ed. James Strachey. London: Hogarth P, 1953.

Freud, Sigmund. "Mourning and Melancholia." *The Standard Edition of the Complete Psychological Works of Sigmund Freud*. 1915. Volume 14. Ed. James Strachey. London: Hogarth P, 1955. 237–258.

Friedman, Thomas. "Foreign Affairs Big Mac." *New York Times*, 8 December 1996.

Friedman, Thomas. *The World Is Flat: A Brief History of the Twenty-First Century*. New York: FSG, 2005.

Grandoni, Dino. "Famed Climate Scientist Wins Million-Dollar Verdict Against Right-Wing Bloggers." *The Washington Post*, 8 February 2024. https://www.washingtonpost.com/climate-environment/2024/02/08/michael-mann-bloggers-defamation-trial/#:~:text=Michael%20Mann%2C%20a%20prominent%20climate,victory%20for%20the%20outspoken%20researcher.

Han, Byung-Chul. *The Burnout Society*. Trans. Erik Butler. Stanford: Stanford UP, 2015.

Hegel, G.W.F. *The Phenomenology of Spirit*. 1807. Trans. A.V. Miller. Oxford: Oxford UP, 1977.

Hook, Derek. "Pilfered Pleasure: On Racism as 'The Theft of Enjoyment.'" *Lacan and Race: Racism, Identity, and Psychoanalytic Theory*. Eds. Sheldon George and Derek Hook. London: Routledge, 2022. 35–50.

Hyppolite, Jean. *Genesis and Structure of the Phenomenology*. 1946. Trans. Samuel Cherniak and John Heckman. Evanston: Northwestern UP, 1974.

James, C.L.R. *Mariners, Renegades and Castaways: The Story of Herman Melville and the World We Live in.* 1953. London: Allison and Busby, 1985.

Klein, Naomi. "Welcome to the Mirror World, Where Nothing Is as It Seems." *The Globe and Mail*, 8 September 2023. https://www.theglobeandmail.com/opinion/article-welcome-to-the-mirror-world-where-nothing-is-as-it-seems/.

Krillic, Alma. "Confinement and *Jouissance* in Herman Melville's 'Bartleby the Scrivener: A Tale of Wall Street.'" In Burnham and Kingsbury, Eds., 59–74.

Krzych, Scott. "Circumstantial Sublimation and Steven Soderbergh's *Ordinary Objects*." *Psychoanalysis, Culture & Society* 23.2 (2017): 123–140.

Lacan, Jacques. *The Seminar of Jacques Lacan: Seminar XI, The Four Fundamental Concepts of Psychoanalysis*. Trans. Alan Sheridan. New York: Norton, 1998.

Langford, Sasha. "The Psychotopology of Climate." In Burnham and Kingsbury, Eds., 97–116.

Lertzman, Renee. *Environmental Melancholia: Psychoanalytic Dimensions of Engagement*. New York: Routledge, 2015.

Mann, Michael E. *The Hockey Stick and the Climate Wars: Dispatches from the Front Lines*. New York: Columbia UP, 2012.

Mann, Michael E. "Science Journals Must Be Unpolluted by Politics." *Wall Street Journal* (31 December 2009). https://www.wsj.com/articles/SB10001424052748703478704574612400823765102.

Milne, Drew. "The Beautiful Soul: From Hegel to Beckett." *Diacritics* 32.1 (Spring 2002): 63–82.

Oreskes, Naomi and Erik M. Conway. *Merchants of Doubt: How a Handful of Scientists Obscured the Truth on Issues from Tobacco Smoke to Global Warming*. New York: Bloomsbury, 2010.

Pippin, Robert. "Back to Hegel?" *Mediations* 26 (Fall/Spring 2012–13). Web. Accessed 23 May 2025.

Porter, M.E. "Marginal Recycling: Place and Informal Recycling in St. John's, Newfoundland." *Local Environment* 20.2 (2015): 149–164.

Ramírez-i-Ollé, Meritxell. "Rhetorical Strategies for Scientific Authority: A Boundary-Work Analysis of 'Climategate.'" *Science as Culture* 24.4 (2015): 384–411.

Riesewieck, Moritz and Hans Block, dirs. *The Cleaners*. Film. Gebrueder Beetz Filmproduktion, 2018.

Ruti, Mari. *Distillations: Theory, Ethics, Affect*. New York: Bloomsbury, 2018.

Ruti, Mari. "The Fall of Fantasies: A Lacanian Reading of Lack." *Journal of the American Psychoanalytic Association* 56.2 (2008): 483–508.

Ruti, Mari. *A World of Fragile Things: Psychoanalysis and the Art of Living*. Albany: SUNY P, 2009.

Steffen, Will, Wendy Broadgate, Lisa Deutsch, Owen Gaffney and Cornelia Ludwig. "The Trajectory of the Anthropocene: The Great Acceleration." *The Anthropocene Review* 2.1 (March 2015): 81–98.

Terada, Rei. *Metaracial: Hegel, Antiblackness, and Political Identity*. Chicago: U of Chicago P, 2023.

"Tiny House Warriors." https://idlenomore.ca/tiny-house-warriors/. Accessed 23 May 2025.

"Wangari Maathai." https://www.nobelprize.org/prizes/peace/2004/maathai/facts/#:~:text=Wangari%20Maathai%20was%20the%20first, her%20home%20country%20of%20Kenya. Accessed 23 May 2025.

Žižek, Slavoj. *Freedom: A Disease Without Cure*. London: Bloomsbury, 2023.

Žižek, Slavoj. *Iraq: The Borrowed Kettle*. London: Verso, 2004.

Žižek, Slavoj. *Living in the End Times*. London: Verso, 2010.

Žižek, Slavoj. "Melancholy and the Act." *Critical Inquiry* 26.4 (Summer 2000): 657–681.

Žižek, Slavoj. *The Parallax View*. Cambridge: MIT P, 2006.

Žižek, Slavoj. *Tarrying with the Negative: Kant, Hegel, and the Critique of Ideology*. Durham: Duke UP, 1993.

Coda

Centrism and the ethical subject

In *The Lacanian Left*, Yannis Stavrakakis argues against the very valorization of the suicidal act, and the politics of the drive, decried by Ruti, pointing to an immanent contradiction in that ethical act, which, Stavrakakis says, "has to be purified from its stress on negativity and lack" (Stavrakakis 121). But he also makes a more fundamental point that the concept of the

> "Lacanian Left" can only be a signifier of its own division, a division which is not to be repressed or disavowed but, instead, highlighted and negotiated again and again as a locus of immense productivity, as the encounter – within theoretical discourse – of the constitutive gap between the symbolic and the real, knowledge and truth, the social and the political.
>
> (Stavrakakis 4–5)

Such a constitutive gap or lack I have already demarcated within or between Lacanian centrism and centrist Lacanians, which is not to speak of "left" and "right" in normative form, neither its post-political disavowal nor its neopopulist acceleration. It is not horseshoe/convergence theory but also not *coincidentia oppositorum*-slash-*Doppelgänger* (Klein), whereby the more social justice aspect of Ruti around gender, queer, and class issues and marking of Lacanian limits with respect to the same: "I say this as someone for whom poverty and social class are not only theoretically, but personally, important" (*SB* 215). But also Ruti's own limits vis-à-vis Puar and Brown: this concatenation, dare one call it a clusterfuck, makes Ruti more left than the Lacanian

DOI: 10.4324/9781003518914-6

centrism: so-called revolutionary Lacanians, what Stavrakakis called *The Lacanian Left*.

How does this concept of centrism help us think of the question of the ethical subject, who, committed to the singularity of their desire, does not give up on their jouissance, sees that there is no big Other? Such a radicality of subjectivity should teach us something about our own political position, especially if we admit that the climate denialist may indeed hold that position. Or is the ethical subject, rather, yet another example of the political uselessness of Lacanian theory, due to, as one hostile critic put it, "an ontological and epistemological restrictiveness derived from its absolutizing of structural topologies which limits its explanatory power by rendering far too many specificities of metropolitan statist societies 'necessary' or 'unavoidable'" (Robinson 356)?

The fungibility of (emancipatory) grief and the contradictions of desire

Ruti's theory of grief and sublimation tells us something different, however. First, by positing that the "two levels of lack – constitutive and circumstantial – are intimately related in the sense that it is often through circumstantial experiences of wounding that we are brought face to face with our constitutive wounding" (*EOO* 131). Perhaps this can be put even more strongly: one only faces one's lack-in-being via such contingent or circumstantial negativity: grief lacks, it is characterized by contradiction, antagonism. The reader will remember that I argued against Ruti's position: in Chapter 2, I said that when our material or concrete existence collapses, we cling to our fantasies (of nation, identity, family) all the more strongly. But Ruti's formulation does work in terms of hewing to Stavrakakis' notion of a productive division or lack, while also answering the critic who wants Lacanian theory to pay more attention to on-the-ground issues. Then, in the decade following *EOO*, Ruti comes to think of the reparative work sublimation and creativity can do for one's constitutive and contingent lack, but not the socioeconomic. Now, that is, the contingent or benchmark trauma – for Ruti, her cancer diagnosis, treatment, and, eventually, death – is separated from the circumstantial lack or negativity of race, class, gender, and sexualities.

But perhaps this limitation of the possibilities of sublimation should not keep us from seeing, in Ruti's work, a politics of emancipatory grief. In *From the Ashes: Grief and Revolution in a World on*

Fire, Sarah Jaffe connects various kinds of grief, from that felt by those who lost family and comrades to COVID-19, to those suffering from climate disasters, and the massive economic dislocations that have, in many ways, contributed to the current neopopulist upswell. "COVID grief is unequally distributed" (Jaffe 201), and so, too, are capacities for enduring it, whether one's desires are firmly attached to capitalist acquisition or to its destruction: for Jaffe, grief ends up being a pathway to solidarity.

I gravitated toward Jaffe's book after hearing her interviewed on the *Ordinary Unhappiness* podcast, in part because she also wrote about her father's death, and how that hit her, and how she then connected to these other social forms of grief – emancipatory grief. Is my concept of the fungibility of grief, or how different griefs hit me in 2023 with Mari's death, then the forest fire season, then my father's death, related to my earlier concept of the clusterfuck, in 2020, when the George Floyd protests, and the climate crisis, and COVID-19 all took place at the same time? Then, in what ways are these fungibilities or clusterfucks related to Ruti's ideas for how different lacks relate, and how to attend to those lacks or trauma? And is this a way of thinking about emancipation?

Another division that we have seen before, perhaps a "locus of immense productivity" (Stavrakakis 4), is that between the two forms of desire, the contradiction in Lacan's theory that Ruti points us to. On the one hand, don't cede your desire; on the other, desire is the desire of the Other. This circle cannot be squared – so the climate denialist who, qua ethical subject refuses to cede their desire, is necessarily ignoring the other desire – the desire of the Other. This assertion of mine may seem tautological, but let me explain as this book winds down. On the one hand, the climate denier does not cede their desire, remains true to the petro-fictions that organize their life: "it often seems impossible," Jaffe acknowledges, "to get people to see beyond a world in which they drive their own car to a certain kind of job that pays the bills for a certain kind of family" (Jaffe 232: here climate denialism is linked to the very forms of marriage and the family from which our queer theorists, Butler or Edelman, opted out). And in so doing, the blue-collar denialist denies or disavows the second imperative of Lacan: that their desire is the desire of the Other: "[a]nd so industrial breadwinner masculinity clings to fossil fuels as the backbone of a system that once created well-paid jobs" (Jaffe 232). There is no grief, then, without an emancipatory grief.

Mari Ruti's late style

This book derives its argument from two conditions: first are the two late-in-life essays in which Mari Ruti offers a new theory of grief and lack. Then, that work arriving just as Ruti herself dies, a situation that takes on world-historical proportions in the forest fire summer of 2023. But what of the belatedness of Ruti's essays? What does it mean, that is, to think about Ruti's texts – and in particular their turn to autotheory – as a "late style"? This concept, "late style," has been worked over by Adorno and Said, in particular (Singh), and can mean a writer or artist's becoming more radical or more restrained in their late period of creativity or intellectual production. In his brief essay on "the very late" Beethoven, Adorno argues that the quality of dissensus or the "torn half" as he would later call it cannot be put down to the composer's impending death but to a final irreconcilability of the artworks' putative content and their formal qualities (Adorno, "Late Style"). In an essay in *The London Review of Books*, Edward Said discusses Adorno's short text at length as well as late works – actually, posthumously published works – by Giuseppe Tomasi di Lampedusa and Constantine Cavafy. Said's reading of these two writers is of less interest here for its content than for his own impending death: "what of the last or late period of life," he asks, "the decay of the body, the onset of ill health (which, in a younger person, brings on the possibility of an untimely end)? These issues, which interest me for obvious personal reasons" (Said, "Thoughts on Late Style"). But the problem here is that what appears to be a form of biographical periodization is also subject to the contingency of the Real, and so what may develop out of the trajectory of one's work over decades (that is, from the past), will, with foreknowledge of one's death, as Ruti had, take on a different temporality, one that is future oriented. Or, perhaps, "no future" oriented. Ruti's two final essays are by definition or periodicity late texts; they are also in a late style insofar as they exemplify her turn to autotheory occasioned, she has said, by her diagnosis. (Actually, they are not, strictly speaking, her final writings or final publications: *The Creative Self*, co-written with Gail Newman, appeared in early 2025 when I was completing this manuscript.) We can, when Mari Ruti's essays are laid against Adorno and Said's texts, think of the first (Adorno's essay) as one about late style but not a belated text; the second (Said's article) both about late style and belated texts (indeed, with a necessary connection); and the third (Ruti's "BB" and "WCNC") as in a late style, a belated pair of texts, but not about belatedness.

References

Adorno, Theodor W. "Late Style in Beethoven." *Essays on Music*. Trans. Susan H. Gillespie. Stanford: Stanford UP, 2002. 564–568.

"Grief, Loss, and Love feat. Sarah Jaffe." *Ordinary Unhappiness* podcast, ep. 64. https://ordinaryunhappiness.buzzsprout.com/2131830/episodes/15594532-64-grief-loss-and-love-feat-sarah-jaffe. Accessed 24 May 2025.

Jaffe, Sarah. *From the Ashes: Grief and Revolution in a World on Fire*. New York: Bold Type, 2024.

Robinson, Andrew. "The Lacanian Left: Psychoanalysis, Theory, Politics." *Contemporary Political Theory* 7 (2008): 351–357.

Ruti, Mari. "The Brokenness of Being: Lacanian Theory and Benchmark Traumas." *Angelaki* 28.6 (November 2023): 123–170.

Ruti, Mari. *Distillations: Theory, Ethics, Affect*. New York: Bloomsbury, 2018.

Ruti, Mari. *The Ethics of Opting Out: Queer Theory's Defiant Subjects*. New York: Columbia UP, 2017.

Ruti, Mari. "When the Cure Is that There Is No Cure: Melancholia, Mourning, Creativity." *Meaningless Suffering: Traumatic Marginalisation and Ethical Responsibility*. Eds. David Goodman and Mookie Manalili. New York: Routledge, 2024. 4–28.

Ruti, Mari and Gail M. Newman. *The Creative Self: Beyond Individualism*. New York: Columbia UP, 2025.

Said, Edward. "Thoughts on Late Style." *The London Review of Books* 26.15 (5 August 2004).

Singh, Surti. "Anxiety, Alienation and Exile: Adorno and Said on Lateness." *Apocalyptic Anxieties: 40th Anniversary Conference of the Institute for the Humanities*. Vancouver: Simon Fraser University, 28 October 2023.

Stavrakakis, Yannis. *The Lacanian Left: Psychoanalysis, Theory, Politics*. Albany: State U of New York P, 2007.

Index

For Product Safety Concerns and Information please contact our EU
representative GPSR@taylorandfrancis.com
Taylor & Francis Verlag GmbH, Kaufingerstraße 24, 80331 München, Germany